TRANSISTOR
CIRCUIT DESIGN
WITH EXPERIMENTS

No. 1875
$21.95

TRANSISTOR
CIRCUIT DESIGN
WITH EXPERIMENTS

DELTON T. HORN

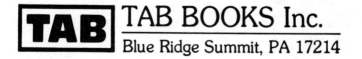

TAB BOOKS Inc.
Blue Ridge Summit, PA 17214

FIRST EDITION

FIRST PRINTING

Copyright © 1985 by TAB BOOKS Inc.
Printed in the United States of America

Reproduction or publication of the content in any manner, without express permission of the publisher, is prohibited. No liability is assumed with respect to the use of the information herein.

Library of Congress Cataloging in Publication Data

Horn, Delton T.
    Transistor circuit design—with experiments.

    Includes index.
    1. Transistor circuits.   I. Title.
TK7871.9.H64   1985   621.3815′30422   85-2653
ISBN 0-8306-0875-3
ISBN 0-8306-1875-9 (pbk.)

Contents

Other TAB books by the Author

Introduction

Building circuits from plans in magazines and books is fun, and it can save you quite a bit of money over buying commercially manufactured equipment. And there's always that glow of pride in using something you made yourself.

You can increase your satisfaction by going one step further and designing your own projects. Knowing how to design circuits can also come in extremely handy when you need a circuit to preform a specific function and can't find a published circuit that will do the job.

In this book I concentrate primarily on the bipolar transistor. Despite the ever increasing use of integrated circuits, the discrete transistor is still an extremely useful device. It is far from obsolete.

No single volume can turn anyone into a full-fledged professional circuit designer, of course. But I believe this book will give you a good head start into design on the hobbyist level. It is also a good starting point for the professional electronic technician who wants to increase his skills by learning the principles of circuit design.

A number of experiments are outlined in the text. It is highly recommended that you actually perform these experiments yourself. No amount of reading can ever take the place of practical, hands-on experience. In addition, you can devise your own experiments by breadboarding the circuits described in the design examples throughout the book. Try changing the various parameters to achieve specific results.

A companion volume to this book on designing IC circuits will soon be published (*Designing IC Circuits . . . with Experiments*—TAB book No. 1925).

Chapter 1

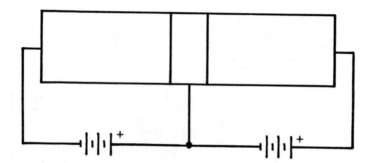

Basic Principles

Early electronic circuits were built around vacuum tubes which, while a miracle of their time, left a lot to be desired. Tubes are bulky and fragile. They use up a lot of power and put out a lot of heat. Moreover, they eventually burn out, reducing the overall reliability of the equipment they are used in.

The field of electronics didn't really start to take off until the development of semiconductors.

What is a semiconductor? Most people have some familiarity with conductors and insulators, but semiconductors sound pretty strange.

Everything is made up of tiny particles called atoms. Atoms, in turn, are made up of tinier subparticles. The number and arrangement of these subparticles are what determine what kind of atom it is.

Many different types of subparticles have been discovered within atoms, but only three need to concern us here. They are the electrons, the protons, and the neutrons. A standard simplified model of an atom is shown in Fig. 1-1. Notice that the protons and neutrons are clumped together in the center (the nucleus) of the atom, while the electrons circle around the outskirts of the atom, like planets orbiting around a sun.

Electricity is the flow of electrons from atom to atom. Some types of atoms (elements) are quite willing to trade off their electrons without a fight. Electricity therefore passes very easily

1

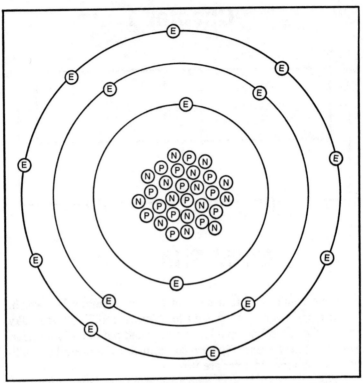

Fig. 1-1. An atom is made up of three primary components—electrons, protons, and neutrons.

through such elements, which are called *conductors*. The metals are usually the best conductor elements. Typical good conductors include copper and silver.

Other elements are very reluctant to give up any of their electrons, or to accept any intruding electrons from other atoms. Such substances oppose the flow of electricity, and are called *insulators* It should be noted that an insulator can conduct if the electrical force is strong enough. There is no such thing as an absolutely perfect insulator. Typical good insulators include rubber and glass.

Semiconductors fall somewhere in the middle. They aren't particularly good conductors, and they aren't particularly good insulators. Well, so what? What good are they?

Actually, we will soon see that semiconductors have some very useful and rather amazing properties.

Two of the most common elements used for semiconductors are germanium and silicon. These are crystalline substances that in their pure form don't have any special electrical properties. In fact, pure germanium or pure silicon are pretty good insulators.

But when a very small impurity of the right type is added to germanium or silicon crystals, the situation changes completely. The amount of impurity is minute—often as small as one part in 10,000,000. That's all it takes.

If a few arsenic atoms are added to a slab of germanium, the arsenic will try to act like germanium and join in the crystalline structure, but it has too many electrons in its outermost ring. The result is a few loose electrons wandering around in the germanium crystal.

Ordinarily atoms and molecules are electrically neutral. Each electron has one unit of negative electrical charge. Each proton has one unit of positive electrical charge. Ordinarily an atom has an equal number of electrons and protons, cancelling each other out, leaving the atom electrically neutral as a whole.

The loose electrons wandering around within our doped germanium crystal represent a wandering local negative charge, but the overall charge of the crystal as a whole is neutral, because the number of electrons exactly equals the number of protons. The loose electrons just can't find a place to "sit down and make themselves comfortable," so they wander aimlessly about, looking for a good place to settle.

Now, let's say an electrical voltage is placed across the crystal. The loose electrons will be drawn to the positive terminal, because there is nothing much to hold them in place within the crystal. So they leave the crystal altogether.

Now there are more protons in the crystal than electrons. The crystal as a whole has a positive charge. It will attract electrons from the negative terminal of the voltage source. This neutralizes the crystal's electrical charge, but the new electrons still can't find anyplace to "sit" so they are drawn out of the crystal by the positive side of the voltage source. This continues as long as the voltage source is applied to the crystal.

The semiconductor material is conducting electricity, but in a somewhat different way than ordinary conductors.

Well, so what? We'll get to that in just a minute.

Using arsenic as an impurity adds extra electrons to the crystalline structure. Therefore, arsenic is called a *donor impurity*. Other donor impurities are antimony, bismuth, and phosphorus.

Alternatively, we could dope the crystal with an impurity with too few electrons to fit the crystalline structure. In other words, there are "holes" where more electrons would fit if they were available. The electrical design of the crystal as a whole is still electrically neutral—the number of electrons equals the number of protons. There are just extra places for electrons to "sit."

The electrons will keep trying to fill up these holes. They sort of play "musical chairs," but it doesn't accomplish much, since there are always more holes than electrons. In a real sense, the holes move around within the crystal, just as the loose electrons did in the earlier version. We now have loose holes. A minute localized positive charge drifts about within the crystal.

Once again, if a voltage source is applied to the semiconductor crystal, current will flow. Electrons will be pulled in from the negative terminal to fill the excess holes. This gives the crystal as a whole a negative electrical charge, which is tapped off by the positive terminal.

Impurities with too few electrons are called *acceptor impurities*. Typical elements used as acceptor impurities are aluminum, boron, gallium, indium, and thallium.

THE PN JUNCTION

As we have seen, there are two types of doped semiconductors. One type has loose electrons (not enough holes), and is called an *N-type* semiconductor. The other type, which is called a *P-type* semiconductor, has loose holes (not enough electrons).

Neither N-type nor P-type semiconductors are particularly exciting or interesting by themselves. But when we form a junction between two different types of semiconductors, some interesting effects and properties start to show up.

It is important to remember that both N-type and P-type semiconductors have both electrons and holes flowing through them. The difference is in which outnumbers the others. In an N-type semiconductor, electrons (negative charge) are the majority carriers, and holes are the minority carriers. In a P-type semiconductor, we have just the opposite—the majority carriers are holes (positive charge), and the minority carriers are electrons.

When no voltage is applied to a PN junction, the carriers (electrons and holes) are more or less randomly distributed, as illustrated in Fig. 1-2.

In Fig. 1-3 we are applying a voltage across the PN junction. The positive voltage terminal is connected to the N-type semiconductor, and the negative voltage terminal is connected to the P-

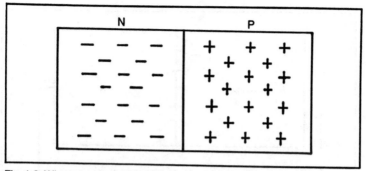

Fig. 1-2. When no voltage is applied to a PN semiconductor junction, the electrons and holes are randomly scattered throughout the material.

type semiconductor. The excess electrons are drawn towards the positive terminal on the N side of the junction. Similarly, the excess holes drift towards the negative terminal on the P side.

All the loose carriers are drawn to the ends of the semiconductor slab. There are virtually no loose electrons or holes near the junction at the middle. Essentially, the junction is the same as if it had never been doped with any impurity. An undoped semiconductor, remember, is an insulator. The result is that no current will flow through the junction. It behaves almost like an open circuit. (There will be a very small amount of current flow due to the minority carriers, but this is small enough that it can be ignored for our purposes here.)

Now, let's see what happens when the polarity of the voltage source is reversed, as shown in Fig. 1-4. Now the negative terminal

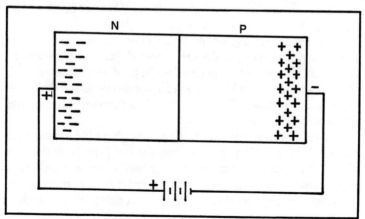

Fig. 1-3. When a PN junction is reverse-biased, current flow is blocked.

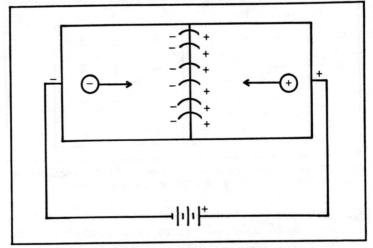

Fig. 1-4. Current can flow through a forward-biased PN junction.

is connected to the N-type semiconductor. Since like charges repel, the loose electrons are forced towards the center of the slab—that is, they move towards the junction. Similarly, the positive voltage terminal is connected to the P-type semiconductor, forcing the loose holes towards the junction.

The loose electrons and holes jump across the junction and neutralize each other. The loose electrons fill the loose holes. Now the P-type semiconductor has more electrons than protons since extra electrons from the N side have filled its holes. This gives the P side a negative electrical charge. At the same time, the N side has more protons than electrons (since it has lost its loose electrons to the P side), so it has a positive charge.

The positive charge on the N side draws more electrons from the negative terminal, while the excess electrons flow from the negatively charged P-type material to the positive voltage terminal. This means the N side is again given loose electrons and the P side again has loose holes. These are forced through the junction, and the process continues.

In other words, current can flow through a PN junction from the N side to the P side, but not from the P side to the N side. An ideal semiconductor junction with no minority carriers would have infinite resistance in one direction, and zero resistance in the other. Minority carriers gum things up a bit, so a practical semiconductor junction will exhibit a very high resistance in one direction, and a very low resistance in the other.

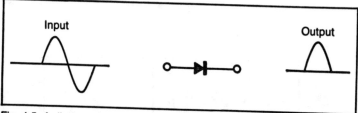

Fig. 1-5. A diode passes current in only one direction.

This one-way current flow is called the *diode effect*. In fact, we have described a semiconductor diode, which is nothing more than a slab of N-type semiconductor and a slab of P-type semiconductor in contact with each other.

If an ac (alternating current) voltage source is placed across a diode, it will be rectified, as illustrated in Fig. 1-5.

The N side of a semiconductor diode is called the *cathode*. The P side is called the *anode*. Current will flow through the diode only when the cathode is made negative with respect to the anode. The standard schematic symbol for a diode is shown in Fig. 1-6. The arrowhead points towards the cathode. Usually the cathode will be marked on the casing of the diode, as in Figs. 1-6B and 1-6C.

THE BIPOLAR TRANSISTOR

The diode effect can be very useful in a large number of applications, but we can do even more if we have a pair of back-to-back PN junctions, as shown in Fig. 1-7. In effect, we have a semiconductor sandwich. This device is called a *transistor*.

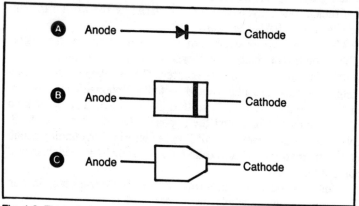

Fig. 1-6. The polarity of a diode is indicated on its housing.

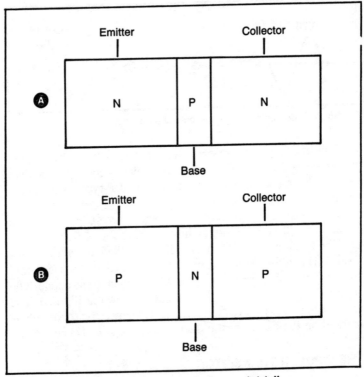

Fig. 1-7. A transistor is like a "semiconductor sandwich."

Specifically, the official name is bipolar transistor, since there are other types of transistors. But usually, when we just say transistor, without specifying the type, we mean bipolar transistor.

Incidentally, if you care about such things, the word transistor comes from TRANSfer resISTOR.

There are two possible combinations for a bipolar transistor. There could be two slabs of N-type material on the outside, with a slab of P-type material in the center. This is called an *NPN* transistor, for reasons which should be fairly obvious. The structure of this type of device is illustrated in Fig. 1-7A. The other kind of bipolar transistor is just the opposite. As shown in Fig. 1-7B, this time we have two slabs of P-type material surrounding a slab of N-type material. This is called a *PNP* transistor, which should come as no surprise.

For now we will concentrate on the NPN transistor in our discussion of how a bipolar transistor works.

Notice that the center P section is much thinner than either

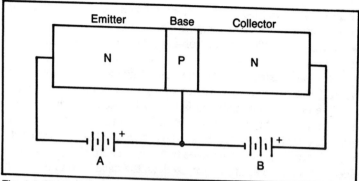

Fig. 1-8. The batteries are arranged to correctly bias the NPN transistor.

of the outside N sections. The P material has fewer holes than either of the N material sections have electrons. We will learn the importance of this shortly.

A bipolar transistor has three leads, each connected to one of the three sections of the semiconductor slab. One of the outer sections is called the emitter, the center section is called the base, and the other outer section is called the collector.

For an NPN transistor, the emitter is made negative with respect to the base, which in turn, is made negative with respect to the collector. In other words, the emitter is the most negative terminal, and the collector is the most positive terminal. The base is somewhere in between. The proper polarity connections for an NPN transistor are illustrated in Fig. 1-8.

In Fig. 1-9, we have broken up the bipolar transistor into two

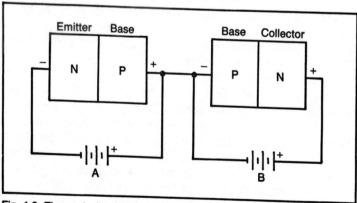

Fig. 1-9. The equivalent circuit of a forward-biased NPN transistor.

separate PN junction diodes. Here we can clearly see that the emitter-base junction is forward-biased (allowing current flow), while the base-collector junction is reverse-biased (blocking current flow).

Now, let's examine what happens within the bipolar transistor, when it is connected to a pair of voltage sources in this manner.

The emitter-base junction is forward-biased. The loose electrons in the emitter section will be forced by the negative voltage terminal over into the base. Since the base is smaller than the emitter, it doesn't have enough holes to accommodate all the invading electrons. Few holes are neutralized, so the base section does not pick up much of a negative charge. Only a few electrons are drawn out through the positive terminal of voltage source A.

There are two voltages affecting the electrons entering the base of the emitter. On the one hand we have voltage A, and on the other we have the series combination of voltages A and B. We will call this series combination AB.

Obviously voltage AB is going to be greater than voltage A. In addition, voltage source B is generally made larger than voltage A. The electrical field produced by voltage AB is going to be considerably stronger than the electrical field produced by voltage A. Most of the electrons will tend to follow the stronger electrical field. A few will go along with the weaker field, but they will be a definite minority.

Most of the electrons entering the base from the emitter pass on through the collector to the positive terminal of voltage source B.

Approximately 5% of the electrons leave the base to the positive terminal of voltage source A, while 95% pass through the collector to the positive terminal of voltage source B.

The voltage of A controls how many electrons enter the base from the emitter. Since most of the electrons from the base pass through the collector, the voltage between the emitter and the base (A) controls the amount of current flowing through the collector.

If you are familiar with vacuum tubes, the sections of a bipolar transistor closely resemble those of a triode tube. The emitter of a transistor is analogous to the cathode of a tube (source of electron flow). The base serves essentially the same function as the tube's grid (control of electron flow). The transistor's collector plays the same basic role as the plate in a tube (destination of electron flow within the device). Not surprisingly, a bipolar transistor can be used in many of the same applications as vacuum tubes.

A PNP transistor works in basically the same way as an NPN

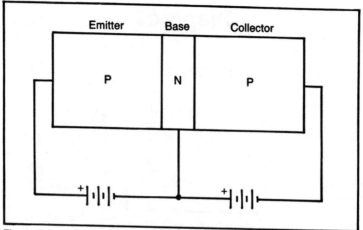

Fig. 1-10. Polarities reversed to forward-bias a PNP transistor.

transistor, except all polarities are reversed, and we are dealing with the flow of holes, instead of the flow of electrons. (The flow of holes is actually the same as the flow of electrons, but in the opposite direction. Conceptually it simplifies things to distinguish between them. To understand the functioning of a PNP transistor, reread the explanation of the NPN transistor and substitute positive for negative, negative for positive, and hole for electron). The correct polarity for forward-biasing a PNP transistor is illustrated in Fig. 1-10.

Great care should be taken to get the voltage polarities right when working with bipolar transistors. If the polarities are reversed, the circuit definitely won't work, and the transistor may be damaged, or completely destroyed. If the base-collector junction is forward-biased, an excessive amount of current will start flowing through the semiconductor crystal. This could cause quite a bit of heat to build up. The delicate semiconductor crystals are very sensitive to temperature, and can be destroyed by too much heat.

Chapter 2

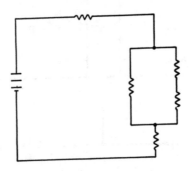

Laws and Formulas
for the Designer

Most of us don't particularly like to tangle with a lot of math. Technical books always run the risk of being dry and dull, and there's nothing like page after page of equations and formulas to make it all very boring and unreadable.

But we might as well face it, mathematical calculations are a major fact of life in electronic circuit design. There's just no getting around it—we're going to have to deal with a number of formulas and equations here. I'll do my best to make it as painless as possible.

This chapter will summarize most of the math we will be using throughout this book. I do not recommend reading this entire chapter at a single sitting, because there is an awful lot of math to take in here. Taken a section at a time, you shouldn't have any major problems. You may find that you are already familiar with much of the material presented here.

If a lot of concentrated math really puts you off, you could temporarily skip ahead to Chapter 3. When you come to where a specific equation is used in the text, you can refer back to the appropriate section in this chapter. Some people may prefer this more broken-up approach. I choose to include all the major formulas we'll be using in a single chapter to make this book more useful for reference. If you need to look up a specific formula at a later date, you'll probably be able to find it in this chapter.

OHM'S LAW

If you've done any work in electronics at all, you should have some familiarity with Ohm's Law. Ohm's Law is a set of simple equations that define the relationship between voltage, current, and resistance in a circuit, or a section of a circuit.

The voltage equals the product of the current (in amperes) multiplied by the resistance (in ohms). That is:

$$E = IR$$

where E is the voltage in volts, I is the current in amperes, and R is the resistance in ohms. E, I, and R are traditionally used to represent these parameters in most electronics equation. Don't worry about why, just learn the convention.

Let's try an example. If the current is 2 amps, and the resistance is 15 ohms, the voltage works out to:

$$E = IR = 2 \times 15 = 30 \text{ volts}$$

That's certainly simple enough.

In most practical electronic circuits, the current will be much smaller than in this example, and the resistance will be much larger. Current is usually given in *milliamps* (mA). One milliamp equals one thousandth of an amp. That is:

$$1 \text{ mA} = 0.001 \text{ amp}$$
$$1 \text{ amp} = 1000 \text{ mA}$$

Occasionally, you may find values given in *microamps* (μA). A microamp is one millionth of an ampere, or one thousandth of a milliamp. That is:

$$1 \text{ μA} = 0.001 \text{ mA} = 0.000001 \text{ amp}$$
$$1 \text{ mA} = 1000 \text{ μA} = 0.001 \text{ amp}$$
$$1 \text{ amp} = 1000 \text{ mA} = 1,000,000 \text{ μA}$$

Resistances are often given in *kilohms*. A kilohm is equal to 1000 ohms. Kilohms is usually abbreviated as K Ω, or just K. The symbol for ohms is Ω. A still larger unit is the *megohm* (M Ω, or M), which equals one million ohms, or 1000 kilohms. That is:

$$1 \text{ Ω} = 0.001K = 0.000001 \text{ M}$$
$$1000 \text{ Ω} = 1K = 0.001 \text{ M}$$
$$1,000,000 \text{ Ω} = 1000K = 1 \text{ M}$$

These prefixes are standardized. Anytime you see "milli", divide by 1000. Anytime you see "micro," divide by 1,000,000.

Anytime you see "kilo," multiply by 1000. Anytime you see "meg," multiply by 1,000,000.

Now, let's return to Ohm's Law for another example. Let's say the current flowing through a circuit is 22 mA, and the circuit resistance is 3.9K. First, we must convert to the standard units (amperes, and ohms):

$$22 \text{ mA} = 22/1000 = 0.022 \text{ amp}$$

and:

$$3.9K = 3.9 \times 1000 = 3900 \text{ amps}$$

Now, we just use Ohm's Law to calculate the voltage:

$$E = IR = 0.022 \times 3900 = 85.8 \text{ volts}$$

Since we are dividing the current by a thousand, and multiplying the resistance by a thousand, we can cancel these operations out, and use milliamps, and kilohms directly:

$$E = 22 \times 3.9 = 85.8 \text{ volts}$$

We get the same result.

The current and the resistance must be in complementary units:

amperes and ohms
milliamps and kilohms
microamps and megohms

Do not try other combinations. If you mix, for example, megohms with milliamps, you will not get a correct answer. When in doubt, convert all values to their base units (amperes, and ohms in this case).

Suppose we know the voltage and the resistance, and need to determine the current. It's a fairly simple matter to rearrange the Ohm's Law formula to solve for current:

$$E = IR$$

$$\frac{E}{R} = \frac{IR}{R}$$

$$\frac{E}{R} = I$$

$$I = \frac{E}{R}$$

As an example, let's assume the voltage is 12 volts, and the resistance is 27K (27,000 ohms). In this case, the current works out to:

$$I = \frac{E}{R} = \frac{12}{27000} = 0.000444 \text{ amps} = 0.444 \text{ mA} = 444 \text{ } \mu A$$

By the same token, we can also rearrange the basic Ohm's Law equation to solve for an unknown resistance:

$$E = IR$$

$$\frac{E}{I} = \frac{IR}{I}$$

$$\frac{E}{I} = R$$

$$R = \frac{E}{I}$$

For our example this time, we'll say the voltage is 9 volts, and the current is 45 mA (0.045 amps). This means, the resistance must be equal to:

$$R = \frac{E}{I} = \frac{9}{0.045} = 200 \text{ ohms}$$

To summarize, Ohm's Law defines the relationship between voltage, current, and resistance. If you know any two of these values, you can calculate the third.

To solve for an unknown voltage, use:

$$E = IR$$

To solve for an unknown current, use:

$$I = \frac{E}{R}$$

To solve for an unknown resistance, use:

$$R = \frac{E}{I}$$

The Ohm's Law equations are most commonly used in dc circuits. In ac circuits things get a bit more complicated. In later sections of this chapter we will work with some formulas to deal with some of the special features of ac values.

POWER

The fourth most important parameter in electrical circuits is the amount of power consumed. The standard unit is the *watt*.

Power, in wattage, equals the product of the voltage, in volts, multiplied by the current, in amps. That is:

$$P = EI$$

If the voltage is one volt, and the current is one ampere, the power consumption will be one watt:

$$P = EI = 1 \times 1 = 1 \text{ watt}$$

As a practical example, let's say we have 100 mA (0.1 amp) of current flowing in a 25-volt circuit. The power consumption would work out to:

$$P = EI = 25 \times 0.1 = 2.5 \text{ watts}$$

In some circumstances it may be convenient to combine the power formula with Ohm's Law. For instance, let's say we know the current and the resistance, and we need to know the power, but we're not too concerned about the voltage. Ohm's Law allows us to derive the voltage from the current and the resistance:

$$E = IR$$

We can substitute this formula for the value of E in the power equation:

$$P = EI = I \times R \times I = I^2R$$

The power equals the current squared, multiplied by the resistance.

As an example, we will say our circuit draws 35 mA (0.035 amps) through 18K (18,000 ohms). What is the power? We just need to plug our values into the modified equation:

$$P = I^2R = (0.035)^2 \times 18000$$
$$= 0.001225 \times 18000 = 22.05 \text{ watts}$$

We can do the same sort of equation rearranging if we know the voltage and the resistance, but not the current, and need to know the power consumption:

$$I = \frac{E}{R}$$

$$P = EI = E \times \frac{E}{R} = \frac{E^2}{R}$$

As an example, if the voltage is 12 volts, and the resistance is 43K (43,000 ohms), the power will work out to:

$$P = \frac{E^2}{R} = \frac{12^2}{43000} = \frac{144}{43000}$$

$$= 0.00335 \text{ watts} = 3.35 \text{ milliwatts}$$

Once again, a milliwatt is equal to one thousandth of a watt.

Let's try another example. The voltage this time is 50 volts, and the resistance is 27K (27,000 ohms). So the power is equal to:

$$P = \frac{E^2}{R} = \frac{50^2}{27000} = \frac{2500}{27000} \quad 0.092 \text{ watts}$$

$$= 0.0926 \text{ watts} = 92.6 \text{ milliwatts}$$

The power equation can also be used in reverse. For example, many household appliances run from the 120-volt ac current. A wattage (power) rating is usually given, but not the current flow. We can solve for the current by rearranging the power equation:

$$P = EI$$

$$\frac{P}{E} = \frac{EI}{E}$$

$$\frac{P}{E} = I$$

If we have a standard household appliance that consumes 17.5 watts, the current flow must be equal to:

$$I = \frac{P}{E} = \frac{17.5}{120} = 0.146 \text{ amps} = 146 \text{ mA}$$

The Ohm's Law and power equations are certainly simple enough. This is fortunate, since they are among the most commonly used calculations in electronics.

RESISTANCES IN COMBINATION

In our simple examples for Ohm's Law and power calculations, we made the assumption that the circuit consisted of a single resistance element, as shown in Fig. 2-1. But what if we have a more complex circuit with multiple resistances, like the one shown in Fig. 2-2? How do we determine the resistance of the circuit as a whole for our calculations?

Luckily, this is not as difficult as you might think. There are two ways a pair of resistance elements can be combined. They may either be connected in series, or in parallel.

Figure 2-3 shows two resistors connected in series. Their values are simply added together to find the total value. If resistor R_a has a value of 3.3K (3300 ohms), and resistor R_b has a value of 6.8K (6800 ohms), what is the value of the series combination?

$$R_t = R_a + R_b = 3300 + 6800$$
$$= 10100 \text{ ohms} = 10.1K$$

This can be extended to cover any number of resistances in

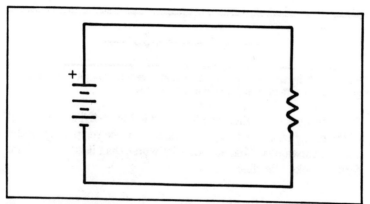

Fig. 2-1. The parameters of this simple one resistance circuit can easily be found with Ohm's Law.

series. For example, if we have five resistances in series, simple addition would give us the total resistance of the combination:

$$R_t = R_a + R_b + R_c + R_d + R_e$$

Parallel combinations of resistances are a bit more complex.

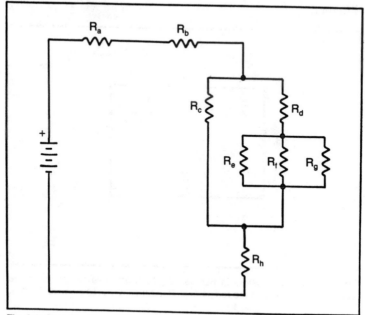

Fig. 2-2. Circuits with multiple resistances are not as obvious in their functioning.

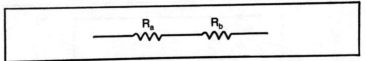

Fig. 2-3. When two resistors are connected in series, their total resistance is equal to the sum of their individual resistances.

Two resistors in parallel are illustrated in Fig. 2-4. The reciprocal of the total resistance is equal to the sum of the reciprocals of the component values. That sounds a lot worse than it is. In equation form it looks like this:

$$\frac{1}{R_t} = \frac{1}{R_a} + \frac{1}{R_b}$$

This equation may alternatively be written like this:

$$R_t = \frac{1}{1/R_a + 1/R_b}$$

This is exactly the same equation. It is just written a little differently.

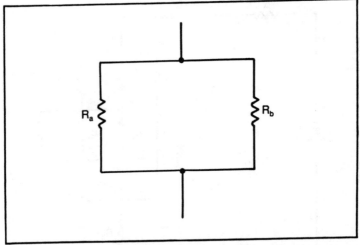

Fig. 2-4. For two resistors in parallel, the total effective resistance is the reciprocal of the sum of the reciprocals of the individual resistances. That is,

$$\frac{1}{R_t} = \frac{1}{R1} + \frac{1}{R2}$$

Let's try an example. If R_a is 3.3K (3300 ohms), and R_b is 6.8K (6800 ohms), we can solve for the total resistance as follows:

$$\frac{1}{R_t} = \frac{1}{R_a} + \frac{1}{R_b} = \frac{1}{3300} + \frac{1}{6800}$$

$$= 0.000303 + 0.0001471 = 0.0004501$$

$$R_t = \frac{1}{0.0004501} \cong 2222 \text{ ohms} \cong 2.2K$$

Remember, you must take the reciprocal of the result of the first part of the equation to get the total resistance.

Once again, this equation can be extended to handle any number of resistance elements in parallel. For example, if we had five resistors in parallel, the equation would become:

$$\frac{1}{R_t} = \frac{1}{R_a} + \frac{1}{R_b} + \frac{1}{R_c} + \frac{1}{R_d} + \frac{1}{R_e}$$

When just two resistors are in parallel, we can use a somewhat different equation:

$$R_t = \frac{R_a \times R_b}{R_a + R_b}$$

We can prove this by repeating our earlier sample problem ($R_a = 3300$ ohms, and $R_b = 6800$ ohms):

$$R_t = \frac{R_a \times R_b}{R_a + R_b} = \frac{3300 \times 6800}{3300 + 6800}$$

$$= \frac{22440000}{10100} \cong 2222 \cong 2.2K$$

We get the same result as with the more general equation. This new equation cannot be used for parallel combinations of more than two resistance elements.

Solving the total circuit resistance for a seemingly complex circuit like the one shown in Fig. 2-2 becomes simply a matter of sequentially performing a number of series and parallel combinations.

We will assume that all eight resistors in Fig. 2-2 have a value of 100 ohms each.

First, we can combine R_a and R_b in series:

$$R_{ab} = R_a + R_b = 100 + 100 = 200 \text{ ohms}$$

We can also solve for the parallel combination of R_e, R_f, and R_g:

$$\frac{1}{R_{efg}} = \frac{1}{R_e} + \frac{1}{R_f} + \frac{1}{R_g}$$

$$= \frac{1}{100} + \frac{1}{100} + \frac{1}{100}$$

$$= 0.01 + 0.01 + 0.01 = 0.03$$

$$R_{efg} = \frac{1}{0.03} = 33.33 \text{ ohms}$$

We can now redraw the diagram, replacing these combinations with single resistance elements, as shown in Fig. 2-5.

The next step is to solve the series combination of R_d and R_{efg}:

$$R_{de} = R_d + R_{efg} = 100 + 33.33$$

$$= 133.33 \text{ ohms}$$

This allows us to again simplify the diagram, as illustrated in Fig. 2-6.

Next, we solve for the parallel combination of R_c and R_{de}:

$$R_{cd} = \frac{R_c \times R_{de}}{R_c + R_{de}} = \frac{100 \times 133.33}{100 + 133.33}$$

$$= \frac{13333}{233.33} \cong 57 \text{ ohms}$$

The diagram is again simplified in Fig. 2-7.

This leaves us with three resistance elements in series, R_{ab},

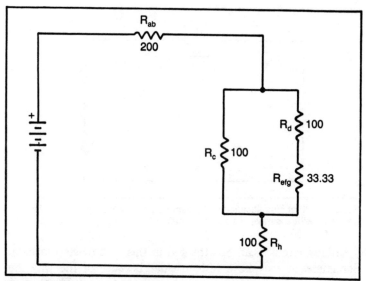

Fig. 2-5. Complex resistance circuit.

R_{cd}, and R_h. It's a simple matter to add their values to get the total circuit resistance:

$$R_t = R_{ab} + R_{cd} + R_h = 200 + 57 + 100$$
$$= 357 \text{ ohms}$$

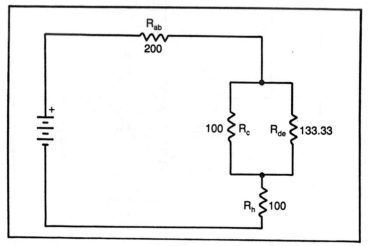

Fig. 2-6. We can simplify the sample circuit of Fig. 2-5 by combining resistances into their effective equivalents.

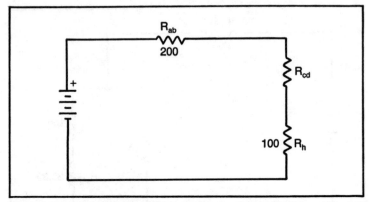

Fig. 2-7. Further simplification of the sample circuit from Fig. 2-5 and Fig. 2-6 leaves us with a simple resistance circuit.

Many circuits can be attacked in this step-by-step fashion. Sometimes it may be a little trickier because of the way the schematic is drawn. It may not be obvious in a circuit like the one shown in Fig. 2-8. Where do you begin combining the resistances here? Almost always, the circuit (or portions of it) may be redrawn into the familiar forms, as shown in Fig. 2-9. Figures 2-8 and 2-9 show exactly the same circuit.

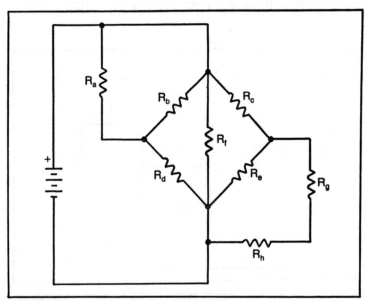

Fig. 2-8. This circuit looks more complicated than it is.

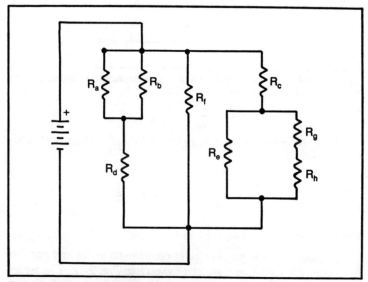

Fig. 2-9. The circuit of Fig. 2-8 can be redrawn to make the series and parallel relationships of the resistors easier to see.

CAPACITANCES IN COMBINATION

Interestingly enough, capacitances can be combined with the same formulas as resistances, but the rules for which equation to use are reversed.

For capacitances in parallel, as shown in Fig. 2-10, the capacitances are added:

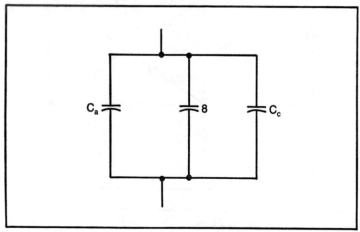

Fig. 2-10. Capacitors connected in parallel are added like resistors in series.

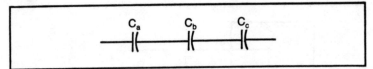

Fig. 2-11. To find the equivalent value of several capacitors in series, use the reciprocal equation, like resistors in parallel.

$$C_t = C_a + C_b + C_c$$

For capacitances in series, as in Fig. 2-11, the reciprocal equation is used:

$$\frac{1}{C_t} = \frac{1}{C_a} + \frac{1}{C_b} + \frac{1}{C_c}$$

KIRCHHOFF'S VOLTAGE LAW

In some circuits, you can't get convenient series and parallel combinations. Consider the circuit shown in Fig. 2-12. The series

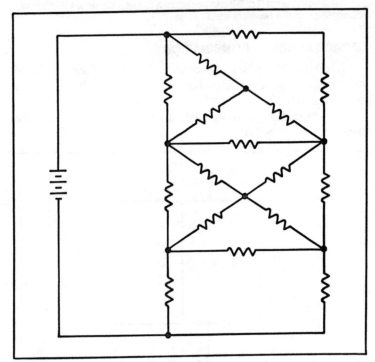

Fig. 2-12. This complex resistance circuit cannot be reduced to simple series/parallel combinations.

and parallel combination rules can't be used to simplify this circuit. How can we analyze this circuit?

One tool which comes in handy in such circuits is Kirchhoff's voltage law.

To discuss Kirchhoff's voltage law, we must first define a *loop*. This is simply any closed conducting path. A loop may include voltage sources (but not current sources), resistances, reactances, and conductors in any series combination. The circuit shown in Fig. 2-13 has three loops, which are illustrated in Fig. 2-14. We only need two to understand the circuit. When using Kirchhoff's voltage law, use the minimum number of loops that contain all the circuit elements. Often there will be several possibilities, and it doesn't matter which you use—you'll get the same results. Therefore, it makes sense to select the loops that will be the easiest to work with.

In Fig. 2-14 the loop currents are shown. A loop current is assumed to flow only within its associated loop. It is a mathematical fiction which may or may not correspond to the real current actually flowing through that portion of the current. The loop current is assumed for mathematical purposes. The equations will work out and give us the correct results, and that's what matters.

In a nutshell, Kirchhoff's voltage law states that *the algebraic sum of all voltage sources in any loop is equal to the algebraic sum of the voltage drops around the loop.* The entire voltage is dropped within the loop. The voltage drops are considered to be caused not just by that loop's current, but by any other loop current flowing through the resistance element in question.

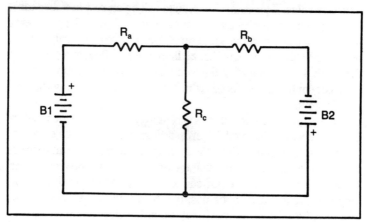

Fig. 2-13. The Kirchhoff's Voltage Law example described in the text uses this circuit.

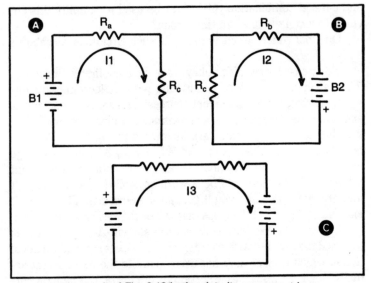

Fig. 2-14. The circuit of Fig. 2-13 broken into its component loops.

Don't let the phrase "algebraic sum" throw you. It just means we'll be dealing with positive and negative values. The choice of signs is arbitrary, but it is important to be consistent.

Select a direction (clockwise, or counterclockwise) for each loop current. It doesn't matter if this relates to the direction of actual current flow or not. If the current is flowing in the opposite direction from the one we selected, the equation will work out to a negative value. The numerical results will be correct in either case.

It usually simplifies matters considerably if all the loop currents are assumed to go in the same direction. I have arbitrarily settled on a standard of assuming that all loop currents flow in a clockwise direction. By always using the same directional convention, I don't ever get confused and forget which way the equations say the current is going.

The sign of the current flow through a resistance element determines the sign of the voltage drop across that resistance element. If the current through the resistance element is in the same direction as the loop current, the voltage drop is positive, otherwise, the voltage drop is negative. Of course, the voltage drop due to the loop current will always be positive, by definition. Voltage drops due to external loop current may be either positive or negative.

If the loop current passes through a voltage source from the negative terminal to the positive terminal, the voltage is given a positive value. If the loop current flows from the positive terminal to the negative terminal, the voltage source takes on a negative value.

Let's put Kirchhoff's voltage law to work analyzing the circuit shown in Fig. 2-13. Loops A and B in Fig. 2-14 contain all the circuit elements, so they will be sufficient for our analysis. We can ignore loop C.

For our example, we will assume the following values for the circuit elements:

$$B1 = 9 \text{ volts}$$
$$B2 = 12 \text{ volts}$$
$$R_a = 10 \text{ ohms}$$
$$R_b = 50 \text{ ohms}$$
$$R_c = 20 \text{ ohms}$$

The circuit is redrawn in Fig. 2-15 with the two loop currents shown.

In loop A we have one voltage source (B1), and two resistance elements (R_a and R_c). Only loop current I1 flows through R_a, so the voltage drop across this element is simply:

$$E_a = I1 \times R_a$$

However, two loop currents (I1 and I2) flow through R_c. Since I2 is flowing in the opposite direction as I1, it is given a negative

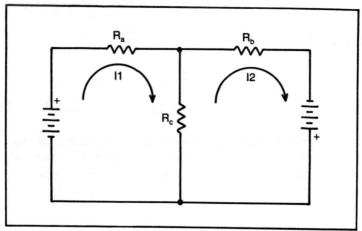

Fig. 2-15. The loop currents through the circuit of Fig. 2-13.

value. The total voltage drop across this resistance element becomes:

$$E_c = (I1 \times R_c) - (I2 \times R_c)$$

According to Kirchhoff's voltage law, the sum of all voltage sources in the loop must be equal to the sum of all voltage drops in the loop. Therefore:

$$B1 = E_a + E_c = (I1 \times R_a) + (I1 \times R_c) - (I2 \times R_c)$$
$$= (I1 \times (R_a + R_c)) - (I2 \times R_c)$$

Plugging in the values from our parts list, we find:

$$9 = (I1 \times (10 + 20)) - (I2 \times 20)$$

$$= (I1 \times 30) - (I2 \times 20)$$

We can do the same thing with loop B:

$$B2 = (I2 \times R_b) + (I2 \times R_c) - (I1 \times R_c)$$

$$= (I2 \times (R_b + R_c)) - (I1 \times R_c)$$

$$12 = (I2 \times (50 + 20)) - (I1 \times 20)$$

$$= (I2 \times 70) - (I1 \times 20)$$

We now have a pair of simultaneous equations with two variables:

$$9 = (I1 \times 30) - (I2 \times 20)$$

$$12 = (I2 \times 70) - (I2 \times 20)$$

There are several methods for solving simultaneous equations. This is probably the easiest. First, rearrange one of the equations, as if solving for one of the variables. We will modify the second equation to give a value for I2:

$$12 = (I2 \times 70) - (I1 \times 20)$$

$$12 + (I1 \times 20) = I2 \times 70$$

$$\frac{12 + (I1 \times 20)}{70} \quad I2$$

Now, we can substitute this formula for I2 in the first equation:

$$9 = (I1 \times 30) - (I2 \times 20)$$

$$= (I1 \times 30) - \left(\frac{12 + (I1 \times 20)}{70} \times 20 \right)$$

$$= (I1 \times 30) - \left(\frac{(12 \times 20) + (I1 \times 20 \times 20)}{70} \right)$$

$$= (I1 \times 30) - \left(\frac{240}{70} + \frac{I1 \times 400}{70} \right)$$

$$= (I1 \times 30) - (3.43 + (I1 \times 5.71))$$

$$= (I1 \times 30) - 3.43 - (I1 \times 5.71)$$

$$= (I1 \times (30 - 5.71)) - 3.43$$

$$= (I1 \times 24.29) - 3.43 = 9$$

Since we have only one unknown variable in this equation now, we can rearrange the equation to solve for the unknown value:

$$9 = (I1 \times 24.29) - 3.43$$

$$9 + 3.43 = (I1 \times 24.29) - 3.43 + 3.43$$

$$\frac{12.43}{24.29} = \frac{I1 \times 24.29}{24.29}$$

$$0.51 = I1$$

Loop current I1 equals approximately 0.51 amp.

Now that we know the value of I1, we can use our modified formula to solve for I2:

$$I2 = \frac{12 + (I1 \times 20)}{70} = \frac{12 + (0.51 \times 20)}{70}$$

$$= \frac{12 + 10.24}{70} = \frac{22.24}{70}$$

$$= 0.32 \text{ amps}$$

Next, we can use these current values to find the voltage drops across each of the resistance elements.

Resistor R_a is only affected by current I1, so:

$$E_a = I1 \times R_a = 0.51 \times 10 = 5.1 \text{ volts}$$

Resistor R_c, however, is affected by both the loop currents, so the voltage drop is slightly more complex:

$$E_c = (I1 \times R_c) - (I2 \times R_c)$$

$$= (0.51 \times 20) - (0.32 \times 20)$$

$$= 10.2 - 6.4 = 3.8 \text{ volts}$$

Finally, R_b is affected only by loop current I2, so its voltage drop works out to:

$$E_b = I2 \times R_b = 0.32 \times 50 = 16 \text{ volts}$$

Wait a minute! How can more voltage be dropped than exists in the loop. Loop B only contains B2 which puts out 12 volts. The answer lies in our sign conventions. I1 runs counter to I2, so for Loop B, the voltage drop across this component is:

$$E_c 2 = (I2 \times R_c) - (I1 \times R_c)$$

$$= (0.32 \times 20) - (0.51 \times 20)$$

$$= 6.4 - 10.2 = -3.8$$

B1 should equal the sum of the voltage drops in loop A (E_a and E_c), and B2 should equal the sum of the voltage drops in loop B; (E_b and $E_c 2$):

$$B1 = 9 = E_a + E_c = 5.1 + 3.8$$

$$= 8.9 \text{ volts}$$

$$B2 = 12 = E_b + E_c2 = 16 - 3.8$$

$$= 12.2 \text{ volts}$$

The slight differences are due to cumulative roundoff errors in the calculations. Our results are close enough for most practical purposes.

Let's try another example with the same circuit (Fig. 2-13). This time we will assume the following component values:

$$B1 = 24 \text{ volts}$$
$$B2 = 6 \text{ volts}$$
$$R_a = 10000 \text{ ohms (10K)}$$
$$R_b = 47000 \text{ ohms (47K)}$$
$$R_c = 180000 \text{ ohms (180K)}$$

Our simultaneous equations are:

$$B1 = 24 = (I1 \times R_a) + (I1 \times R_c) - (I2 \times R_c)$$

$$= (I1 \times (R_a + R_c)) - (I2 \times R_c)$$

$$= (I1 \times (1000 + 180000)) - (I2 \times 180000)$$

$$= (I1 \times 190000) - (I2 \times 180000) = 24 = B1$$

and

$$B2 = 6 = (I2 \times R_b) + (I2 \times R_c) - (I1 \times R_c)$$

$$= (I2 \times (R_b + R_c)) - (I1 \times R_c)$$

$$= (I2 \times (47000 + 180000)) - (I1 \times 180000)$$

$$= (I2 \times 227000) - (I1 \times 180000) = 6 = B2$$

We rewrite equation 2 for I2:

$$I2 = \frac{6 + (I1 \times 180000)}{227000}$$

Replacing this factor in equation 1, we can solve for I1:

$$24 = (I1 \times 190000) - \left(\frac{6 + (I1 \times 180000)}{227000} \times 180000 \right)$$

$$= (I1 \times 190000) - \left(\frac{6 \times 180000}{227000} + \frac{I1 \times 180000 \times 180000}{227000} \right)$$

$$= (I1 \times 190000) - \left(\frac{1080000}{227000} + \frac{I1 \times 32400000000}{227000} \right)$$

$$= (I1 \times 190000) - 4.76 - (I1 \times 142731)$$

$$= 24 = (I1 \times 47269) - 4.76 = 24$$

$$I1 \times 47269 = 28.76$$

$$I1 = 0.0006 \text{ amp} = 0.6 \text{ mA}$$

We can now find I2:

$$I2 = \frac{6 + (I1 \times 18000)}{227000} = \frac{6 + (0.0006 \times 180000)}{227000}$$

$$= \frac{6 + 108}{227000} = \frac{114}{227000} = 0.0005 \text{ amp} = 0.5 \text{ mA}$$

The voltage drops in loop A work out to:

$$E_a = I1 \times R_a = 0.0006 \times 10000 = 6 \text{ volts}$$

$$E_c = (I1 \times R_c) - (I2 \times R_c)$$

$$= (0.0006 \times 180000) - (0.0005 \times 18000)$$

$$= 108 - 90 = 18 \text{ volts}$$

$$B1 = 24 = E_a + E_c = 6 + 18 = 24 \text{ volts}$$

Similarly, the voltage drops in loop B work out to:

$$E_b = I2 \times R_b = 0.0005 \times 47000 = 23.5 \text{ volts}$$

$$E_c 2 = (I2 \times R_c) - (I1 \times R_c)$$

$$= (0.0005 \times 180000) - (0.0006 \times 180000)$$

$$= 90 - 108 = -18 \text{ volts}$$

$$B2 = 6 = E_b + E_c 2 = 23.5 - 18 = 5.5 \text{ volts}$$

Again, we had a half-volt roundoff error.

Any circuit can be analyzed in this fashion, although the exact equations will vary somewhat, depending on the number of voltage sources and resistance elements within each loop, and the number of loops in the circuit.

KIRCHHOFF'S CURRENT LAW

Besides his voltage law, Kirchhoff also came up with a law for analyzing current in complex circuits. Remember that the loop currents we dealt with in Kirchhoff's voltage law were mathematical fictions which may or may not correspond to the actual current flowing through the components.

To deal with actual currents, rather than the mathematical fictions of Kirchhoff's voltage law, we use Kirchhoff's current law.

Again, we have to start with a definition. A *node* is a connection point between two or more conductors. The nodes in our sample circuit are indicated in Fig. 2-16.

According to Kirchhoff's current law, *the amount of current flowing into a node always exactly equals the current flowing out of that node,* which certainly makes sense, when you think about it. In more mathematical terms, the algebraic sum of currents through a node is zero.

Current flowing into a node is assumed to be positive. Current flowing out of a node is assumed to be negative.

For voltage drops across resistance elements, the terminal where the current enters is assumed to be at a higher potential than the terminal where the current exits.

If a circuit has N nodes, we will need to examine N-1 to com-

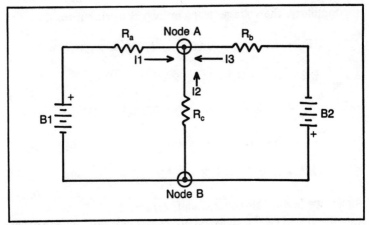

Fig. 2-16. Kirchhoff's Current Law requires that you identify the nodes in the circuit to be analyzed.

pletely analyze the circuit. The required number of node equations is always one less than the total number of nodes.

In our sample circuit of Fig. 2-16, we only have two nodes, so we only need one to solve the circuit.

There are three current paths into node A. These are marked in Fig. 2-16 as I1, I2, and I3. According to Kirchhoff's current law, the algebraic sum of these currents must be equal to zero. That is:

$$I1 + I2 + I3 = 0$$

This won't get us much, until we relate the currents to the voltages and resistances in the circuit.

Current I1 flows through resistor R_a. It is equal to the voltage drop across R_a divided by the resistance of R_a (Ohm's Law—I = E/R).

Now, the voltage drop across R_a must be equal to the voltage going into the resistance element at the positive terminal (which is B1 in this case) minus the voltage at the negative terminal of the resistance element, which we will call E_a. The current direction of I2 means node A is less positive (more negative) than node B, so voltage E_a takes on a negative sign. Putting this together, we can create an Ohm's Law equation for current I1:

$$I1 = \frac{B1 - (-E_a)}{R_a}$$

The two negative signs in front of E_a cancel out, leaving:

$$I1 = \frac{B1 + E_a}{R_a}$$

Current I2 is defined by the voltage drop across R_c. This is simply equal to E_a, so:

$$I2 = \frac{E_a}{R_c}$$

Finally, R3 is determined by the voltage drop across R_b. The input voltage is B2, and the output voltage is E_a. B2 is negative because of the battery polarity. E_a is negative because of the direction of the I2 current flow. This makes I3 equal to:

$$I3 = \frac{-B2 - (-E_a)}{R_b} = \frac{-B2 + E_a}{R_b}$$

$$= \frac{E_a - B2}{R_b}$$

Next, we can substitute these formulas in the original node equation:

$$I1 + I2 + I3 = 0$$

$$\frac{B1 + E_a}{R_a} + \frac{E_a}{R_c} + \frac{E_a - B2}{R_b} = 0$$

We can simplify and rearrange the equation like this:

$$\frac{B1}{R_a} + \frac{E_a}{R_a} + \frac{E_a}{R_c} + \frac{E_a}{R_b} - \frac{B2}{R_b} = 0$$

$$\frac{E_a}{R_a} + \frac{E_a}{R_b} + \frac{E_a}{R_c} + \frac{B1}{R_a} - \frac{B2}{R_b} = 0$$

$$E_a \times \left(\frac{1}{R_a} + \frac{1}{R_b} + \frac{1}{R_c} \right) = \frac{B2}{R_b} - \frac{B1}{R_a}$$

To go any further, we will need some component values to work with. We will use the same values we used in the first sample problem for Kirchhoff's voltage law. These component values were:

$$
\begin{aligned}
B1 &= 9 \text{ volts} \\
B2 &= 12 \text{ volts} \\
R_a &= 10 \text{ ohms} \\
R_b &= 50 \text{ ohms} \\
R_c &= 20 \text{ ohms}
\end{aligned}
$$

Plugging these values into the equation, we find:

$$E_a \times \left(\frac{1}{10} + \frac{1}{50} + \frac{1}{20} \right) = \frac{12}{50} - \frac{9}{10}$$

$$E_a \times (0.1 + 0.02 + 0.05) = 0.24 - 0.9$$

$$E_a \times 0.17 = -0.66$$

$$E_a = \frac{-0.66}{0.17} = -3.88 \text{ volts}$$

The negative sign simply indicates the polarity is the opposite of the one we assumed.

We can now go back and solve for each of the currents:

$$I1 = \frac{B1 + E_a}{R_a} = \frac{9 + (-3.88)}{10} = \frac{9 - 3.88}{10}$$

$$= \frac{5.12}{10} = 0.512 \text{ amp} = 512 \text{ mA}$$

$$I2 = \frac{E_a}{R_c} = \frac{-3.88}{20} = -0.194 \text{ amp} = -194 \text{ mA}$$

$$I3 = \frac{E_a - B2}{R_b} = \frac{-3.88 - 12}{50} = \frac{-15.88}{50}$$

$$= -0.318 \text{ amp} = -318 \text{ mA}$$

The negative values for I2 and I3 simply indicate that the actual direction of current flow is the opposite of that shown in Fig. 2-16.

Let's double-check our work by plugging these derived current values back into the node equation:

$$I1 + I2 + I3 = 0$$
$$0.512 + (-0.194) + (-0.318)$$
$$= 0.512 - 0.194 - 0.318 = 0$$

Yes. It works.

Of course, other circuits will end up with slightly different equations. The more nodes there are, the more equations you will have to work with.

A MORE COMPLEX EXAMPLE OF KIRCHHOFF'S LAWS

Let's put Kirchhoff's voltage law to work in a somewhat more complicated circuit. The circuit we will be using this time is shown in Fig. 2-17.

We will assume the following component values:

B1 = 12 volts	R_b = 330 ohms	R_e = 470 ohms
B2 = 6 volts	R_c = 47 ohms	R_f = 560 ohms
R_a = 100 ohms	R_d = 220 ohms	R_g = 120 ohms

At least three loops are required to include all the circuit elements in this circuit. The three loops I have selected are shown

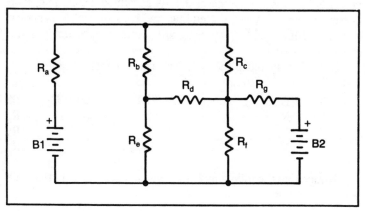

Fig.2-17. This more complex circuit will be analyzed using Kirchhoff's Current Law in the text.

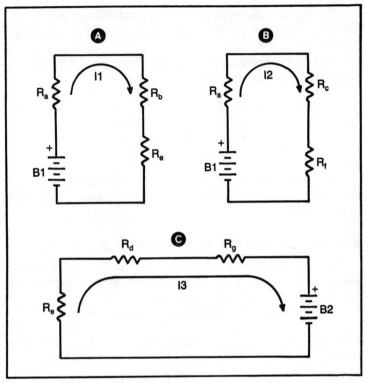

Fig. 2-18. The individual loops in the circuit of Fig. 2-17, separated for analysis.

in Fig. 2-18. In Fig. 2-19, the loops are reassembled with the loop currents indicated. Loop current I1 is shown as a solid line, loop current I2 is shown as a dotted line, and loop current I3 is shown as a dotted line. All loop currents are assumed to be flowing in a clockwise direction.

Each loop includes the following components:

Loop A—	B1	Loop B—	B1	Loop C—	B2
	R_a		R_a		R_e
	R_b		R_c		R_d
	R_e		R_f		R_g

Notice that the following components appear in more than one loop:

$$B1 \quad R_a \quad R_e$$

40

First, we will analyze this circuit with Kirchhoff's voltage law.

Two loop currents (I1 and I2) flow through R_a. Both these currents are flowing in the same direction, so they add, making the voltage drop across this resistance element equal to:

$$E_a = (I1 \times R_a) + (I2 \times R_a)$$

Only loop current I1 flows through R_b, so the equation for the voltage drop across this component is simply:

$$E_b = I1 \times R_b$$

Two loop currents flow through R_e. This time, the currents are flowing in opposite directions, so the voltage drop across this resistance element becomes:

$$E_e = (I1 \times R_e) - (I3 \times R_e)$$

Kirchhoff's voltage law tells us that all voltage sources in a loop must equal the algebraic sum of all voltage drops in that loop. In loop A we have one voltage source (B1), and three voltage drops (E_a, E_b, E_e). Putting them all together we find:

$$B1 = E_a + E_b + E_e = (I1 \times R_a) + (I2 \times R_a) +$$
$$(I1 \times R_b) + (I1 \times R_e) - (I3 \times R_e)$$
$$= (I1 \times (R_a + R_b + R_e)) + (I2 \times R_a) - (I3 \times R_e)$$

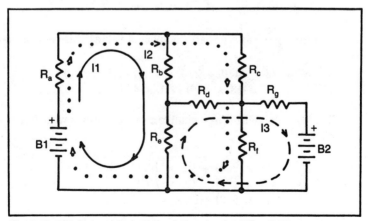

Fig. 2-19. The loop currents through the circuit of Fig. 2-17.

Plugging in the component values from our parts list, this becomes:

$$12 = (I1 \times (100 + 330 + 470)) + (I2 \times 100) - (I3 \times 470)$$
$$= (I1 \times 900) + (I2 \times 100) - (I3 \times 470)$$

Moving on to loop B, we find:

$$E_a = (I1 \times R_a) + (I2 \times R_a)$$
$$E_c = I2 \times R_c$$
$$E_f = I2 \times R_f$$
$$B1 = E_a + E_c + E_f = (I1 \times R_a) + (I2 \times R_a) +$$
$$(I2 \times R_c) + (I2 \times R_f) =$$
$$(I1 \times R_a) + (I2 \times (R_a + R_c + R_f))$$

Substituting our component values, we get:

$$12 = (I1 \times 100) + (I2 \times (100 + 47 + 560))$$
$$= (I1 \times 100) + (I2 \times 707)$$

Finally, for loop C we get:

$$E_e = (I3 \times R_e) - (I1 \times R_e)$$
$$E_d = I3 \times R_d$$
$$E_g = I3 \times R_g$$
$$B2 = E_e + E_d + E_g = (I3 \times R_e) - (I1 \times R_e) +$$
$$(I3 \times R_d) + (I3 \times R_g)$$
$$= (I3 \times (R_e + R_d + R_g)) - (I1 \times R_e)$$

Substituting the appropriate component values, this becomes:

$$6 = (I3 \times (470 + 220 + 120)) - (I1 \times 470)$$
$$= (I3 \times 810) - (I1 \times 470)$$

This leaves us with three simultaneous equations with three variables:

$$I2 = (I1 \times 900) + (I2 \times 100) - (I3 \times 470)$$
$$I2 = (I1 \times 100) + (I2 \times 707)$$
$$6 = (I3 \times 810) - (I1 \times 470)$$

Since the second equation has only the I1 and I2 factors (but not I3), we can rearrange this equation to define I2 in terms of I1:

$$I2 \times 707 = 12 - (I1 \times 100)$$

$$I2 = \frac{12 - (I1 \times 100)}{707}$$

Similarly, the third equation can be rewritten to define I3 in terms of I1:

$$I3 \times 810 = 6 + (I1 \times 470)$$

$$I3 = \frac{6 + (I1 \times 470)}{810}$$

We can now plug these formulas into the first equation, to give us a single unknown variable (I1):

$$12 = (I1 \times 900) + \left(\frac{12 - (I1 \times 100)}{707} \right) \times 100$$

$$- \left(\frac{6 + (I1 \times 470)}{810} \right) \times 470$$

$$= (I1 \times 900) + \left(\frac{12 \times 100}{707} - \frac{(I1 \times 100) \times 100}{707} \right)$$

$$- \left(\frac{6 \times 470}{810} + \frac{(I1 \times 470) \times 470)}{810} \right)$$

$$= (I1 \times 900) + \frac{1200}{707} - \frac{I1 \times 10000}{707} - \frac{2820}{810}$$

$$- \frac{I1 \times 220900}{810}$$

$$= (I1 \times 900) + 1.70 - (I1 \times 1.41) - 3.48$$

$$- (I1 \times 272.72)$$

$$= (I1 \times (900 - 1.41 - 272.72)) + 1.70 - 3.48 =$$

$$(I1 \times 625.87) - 1.78 = 12$$

$$I1 \times 625.87 = 12 + 1.78$$

$$I1 = \frac{13.78}{625.87} = 0.022 \text{ amp} = 22 \text{ mA}$$

This derived value for I1 allows us to solve for I2 and I3:

$$I2 = \frac{12 - (I1 \times 100)}{707} = \frac{12 - (0.022 \times 100)}{707}$$

$$= \frac{12 - 2.2}{707} = \frac{9.8}{707} = 0.014 \text{ amp} = 14 \text{ mA}$$

$$I3 = \frac{6 + (I1 \times 470)}{810} = \frac{6 + (0.022 \times 470)}{810}$$

$$= \frac{6 + 10.34}{810} = \frac{16.34}{810} = 0.020 \text{ amp} = 20 \text{ mA}$$

Now, we can calculate our individual voltage drops across each resistance element:

$$E_a = (I1 \times R_a) + (I2 \times R_a) = (0.022 \times 100) +$$

$$(0.014 \times 100) = 2.2 + 1.4 = 3.6 \text{ volts}$$

$$E_b = I1 \times R_b = 0.022 \times 330 = 7.26 \text{ volts}$$

$$E_c = I2 \times R_c = 0.014 \times 47 = 0.658 \text{ volts}$$

$$E_d = I3 \times R_d = 0.02 \times 220 = 4.4 \text{ volts}$$

$$E_e \text{ (for loop A)} = (I1 \times R_e) - (I3 \times R_e) =$$

$$(0.022 \times 470) - (0.02 \times 470) =$$

$$10.34 - 9.4 = 0.94 \text{ volt}$$

$$E_e \text{ (for loop C)} = (I3 \times R_e) - (I1 \times R_e) =$$

$$= (0.02 \times 470) - (0.022 \times 470)$$

$$= 9.4 - 10.34 = -0.94 \text{ volt}$$

$$E_f = I2 \times R_f = 0.014 \times 560 = 7.84 \text{ volts}$$

$$E_g = I3 \times R_g = 0.02 \times 120 = 2.4 \text{ volts}$$

We can check these voltage drop values by plugging them into the Kirchhoff's voltage law equations:

$$B1 = E_a + E_b + E_e \qquad \text{(Loop A)}$$

$$12 = 3.6 + 7.26 + 0.94 = 11.8 \text{ volts}$$

$$B1 = E_a + E_c + E_f \qquad \text{(Loop B)}$$

$$12 = 3.6 + 0.658 + 7.84 = 12.098 \text{ volts}$$

$$B2 = E_e + E_d + E_g \qquad \text{(Loop C)}$$

$$6 = -0.94 + 4.4 + 2.4 = 5.86 \text{ volts}$$

The small errors are due to rounding off intermediate values during the calculations.

The same circuit could be analyzed via Kirchhoff's current law, but since we would have to solve for four nodes, it would take up considerably more space than we can afford to give it here.

Some circuits will be more convenient to analyze with Kirchhoff's voltage law. Others will be more suitable for Kirchhoff's current law calculations.

PEAK, AVERAGE, AND RMS VALUES

Direct current values are straightforward and reasonably easy to work with. With dc, 10 volts is 10 volts, and that's all there is to that. Alternating current values are a bit more untidy. There are several ways to measure them, because they keep changing from instant to instant.

The most obvious measurement is probably also the most meaningless—the instantaneous value. It's going to be changing in a tiny fraction of a second anyway, so who cares? Besides, it is

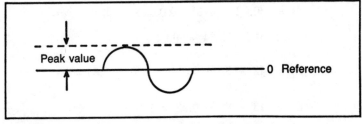

Fig. 2-20. The peak value of an ac signal is the distance between the reference (0) level and the maximum positive value.

difficult to measure, which is just as well.

The next most obvious way to measure an ac signal is to measure its maximum, or peak value. This is useful in some cases, since we often need to know the highest voltage or current a component will be exposed to, but it doesn't really fit in well with most calculations. An ac signal's peak value is indicated in Fig. 2-20.

Closely related to the peak value is the peak-to-peak value, which is illustrated in Fig. 2-21. Since most ac signals are symmetrical around a zero reference level (which may, or may not correspond to actual ground level zero), the peak-to-peak value gives us the full excursion range of the signal. For symmetrical waveforms, the peak-to-peak value is equal to two times the peak value.

Notice that for most waveforms, especially the most common, the sine wave, the peak levels are only briefly reached. Most of the signal is considerably less than the peak values. This is why the peak and peak-to-peak values don't really tell us very much about the signal.

It would seem logical to take an average of a number of instantaneous values throughout an ac cycle, but there is a problem here. Since the signal is usually symmetrical above and below zero, it will always work out to an average of zero for standard waveforms, like the sine wave.

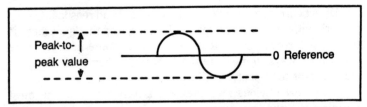

Fig. 2-21. The peak-to-peak value of an ac signal is the distance between the maximum negative level, and the maximum positive level.

Generally, when we speak of the average value of an ac signal, we are referring to an average of only the positive portion of the waveform. The average of the negative portion of the waveform will almost always have the same value, but with the polarity reversed, so this is a reasonable compromise.

In most electronics design work, we will be concerned primarily with sine waves. Therefore, we can resort to some standard formulas. Note that these formulas will probably not be valid for any other waveshape.

For a sine wave, we can derive the average value from the peak value with this equation:

$$AVERAGE = 0.636 \times PEAK$$

Or, working in the reverse direction:

$$PEAK = 1.57 \times AVERAGE$$

The average value is helpful, but it does not correspond to an equivalent dc value, so Ohm's Law won't work. Ohm's Law is such a useful tool that we'd rather not lose it if there's any way we can hang on to it. Fortunately, there is a way.

The voltage drop across a resistor is due to some of the electrical energy being wasted as heat. If an ac signal through a given resistance burns off the same amount of heat as a dc signal through the same resistance, they can be assumed to be equivalent. A method of giving us such ac-dc equivalents is the *Root-Mean-Square* method. This is usually shortened to RMS.

The mathematical derivation for the RMS method is quite complex, and fortunately, it is entirely avoidable for our purposes. The RMS value can be derived from the average value with this simple formula:

$$RMS = 1.11 \times AVERAGE$$

Or, from the peak value, we can get the RMS value with this equation:

$$RMS = 0.707 \times PEAK$$

Of course, we can also convert RMS values back in the other direction:

$$\text{PEAK} = 1.41 \times \text{RMS}$$
$$\text{AVERAGE} = 0.9 \times \text{RMS}$$

Generally, if the form of measurement is not specified, ac values are usually given in RMS units, because this allows the dc equations to be used. However, be forewarned, there are exceptions. You are bound to come across a few unmarked average or peak values from time to time.

Alternating current voltmeters usually measure RMS sine waves. They will not be accurate for other waveshapes.

For your convenience, here is a summary of the standard ac conversion formulas. Remember, these equations are valid for sine waves only:

$$\text{RMS} = 0.707 \times \text{PEAK}$$

$$\text{RMS} = 1.11 \times \text{AVERAGE}$$

$$\text{AVERAGE} = 0.9 \times \text{RMS}$$

$$\text{AVERAGE} = 0.636 \times \text{PEAK}$$

$$\text{PEAK} = 1.41 \times \text{RMS}$$

$$\text{PEAK} = 1.57 \times \text{AVERAGE}$$

$$\text{PEAK} = 0.5 \times \text{PEAK-TO-PEAK}$$

$$\text{PEAK-TO-PEAK} = 2 \times \text{PEAK}$$

REACTANCE AND IMPEDANCE

Resistance is another simple dc property that gets complicated when we start dealing with ac signals. There are several different kinds of ac resistance. They are capacitive reactance, inductive reactance,and impedance.

The major complication with ac resistance is that it varies with the frequency of the ac signal. Dc resistance, of course, is constant. One hundred ohms, for example, is 100 ohms, and that's that. With an ac circuit, however, a component with a nominal impedance of 100 ohms may actually fluctuate between 30 ohms and 675 ohms.

We will start with capacitive reactance. This is the ac resistance of a purely capacitive component (real world capacitors will also exhibit some inductive reactance and dc resistance). The formula for determining the capacitive reactance is:

$$X_c = \frac{1}{2 \pi FC}$$

where X_c is the capacitive reactance in ohms, F is the applied frequency in hertz, and C is the capacitance in farads. The symbol π is pi, a mathematical constant with a value of approximately 3.14. Therefore, 2π is equal to about 6.28, so the equation can be written like this:

$$X_c = \frac{1}{6.28FC}$$

Let's say we are trying to determine the capacitive reactance of a capacitor with a value of $0.2\mu F$ (0.0000002 farad). What is the capacitive reactance? We have no way of knowing unless we define the frequency of the applied signal. If the frequency is 100 Hz, the capacitive reactance equals:

$$X_c = \frac{1}{6.28 \times 100 \times 0.0000002} = \frac{1}{0.0001257}$$

$$= 7958 \text{ ohms}$$

If we raise the frequency to 500 ohms, the capacitive reactance becomes:

$$X_c = \frac{1}{6.28 \times 500 \times 0.0000002} = \frac{1}{0.0006283}$$

$$= 1592 \text{ ohms}$$

As the frequency increases, the capacitive reactance decreases.

If we keep the frequency at 500 Hz, and increase the capacitance to $0.75 \mu F$ (0.00000075 farad), the capacitance reactance is changed to:

$$X_c = \frac{1}{6.28 \times 500 \times 0.00000075} = \frac{1}{0.002356} \cong 424 \text{ ohms}$$

As the capacitance increases, the capacitive reactance decreases.

There is one special case. What is the capacitive reactance for a dc signal? A dc signal may be thought of as having a frequency of 0, so:

$$X_c = \frac{1}{6.28 \times 0 \times 0.00000075} = \frac{1}{0} = \infty$$

Ignoring the effects of inductive reactance and dc resistance, a capacitor exhibits capacitive reactance at dc. That is, it behaves as an open circuit. The value of the capacitor is irrelevant. This is why you will often hear it said that a capacitor passes ac, but blocks dc. Actually, practical capacitors will have some degree of leakage resistance, so they won't behave as a true open circuit at dc, but they will behave as a very high resistance (often several hundred megohms), which will function virtually the same way as an open circuit in many applications.

Inductive reactance, not surprisingly, is the ac resistance of a pure inductor (coil). Again, practical inductors will also exhibit some capacitive reactance and dc resistance, but the inductive reactance will remain the dominant factor.

In many ways, inductive reactance is the opposite of capacitive reactance, as you will soon see.

The formula for inductive reactance is:

$$X_L = 2 \pi FL$$

where X_L is the inductive reactance in ohms, F is the frequency in hertz, and L is the inductance in henries. Once again, 2π is approximately equal to 6.28, of course.

Let's say we have a 100 mH (0.1 henry) coil. If a 100 Hz ac signal is applied, the inductive reactance is equal to:

$$X_L = 6.28 \times 100 \times 0.1 \cong 63 \text{ ohms}$$

Increasing the frequency to 500 Hz changes the inductive reactance to:

$$X_L = 6.28 \times 500 \times 0.1 \cong 314 \text{ ohms}$$

Increasing the frequency increases the inductive reactance.

If we leave the frequency at 500 Hz, and increase the inductance to 330 mH (0.33 henry), the inductive reactance becomes:

$$X_L = 6.28 \times 500 \times 0.33 = 1037 \text{ ohms}$$

Increasing the inductance increases the inductive reactance.

Notice that inductive reactance to changes in frequency and/or inductance in just the opposite way as capacitive reactance responds to changes in frequency and/or capacitance.

Now, let's take the special case of a dc signal being applied to an inductor. What is the inductive reactance in this case?

$$X_L = 6.28 \times 0 \times 0.33 = 0$$

Ignoring the effects of capacitive reactance and dc resistance, the inductive reactance is zero at dc. A perfect inductor would behave like a direct short circuit at dc.

An inductance effectively blocks high frequency ac, and passes dc. Just the opposite of a capacitance.

Ohm's Law will work with reactance (either capacitive, or inductive) in the same way it does with dc resistance, but you must remember that the results are accurate only for a specific frequency.

Let's try a couple of examples. A few pages ago we determined that a $0.2\mu F$ capacitor has a capacitive reactance of approximately 7958 ohms at 100 Hz, and about 1591 ohms at 500 Hz. Assuming the ac voltage is 120 volts, what is the current at each of these frequencies? We simply use the current form of Ohm's Law, substituting capacitive reactance (X_c) for dc resistance (R):

$$I = \frac{E}{X_c}$$

$$I = \frac{120}{7958} = 0.015 \text{ amp} = 15 \text{ mA at 100 Hz.}$$

$$I = \frac{120}{1591} = 0.075 \text{ amp} = 75 \text{ mA at 500 Hz.}$$

Practical circuits include a combination of capacitive reactance, inductive reactance, and dc resistance. The total ac resistance at a specific frequency is called the *impedance*. Remember that impedance, like reactance, is a *frequency specific* value. It will change with the input frequency.

Capacitive and inductive reactances are out of phase with each other, so they cannot simply be added. The formula for impedance is:

$$Z = \sqrt{R^2 = (X_L - X_c)^2}$$

where Z is the impedance, R is the dc resistance, X_L is the inductive reactance, and X_c is the capacitive reactance. All values are in ohms.

The circuit shown in Fig. 2-22 includes an ac voltage source, a capacitor, a coil, and a resistor. We will assume that each of the components is perfect—that is, the capacitor is purely capacitive (no resistive or inductive elements), the resistor is purely resistive, and the coil is purely inductive. Practical components will exhibit some dc resistance and both capacitive and inductive reactance. Fortunately, the leakage values (dc resistance and inductive reactance in a capacitor, for instance) are usually so small they don't have any noticeable effect on the circuit, so they can be ignored.

For our sample problems, we will assume the following component values:

ac voltage = 120 volts
R = 2700 ohms (2.7K)
C = 0.22 μF (0.00000022 farad)
L = 150 mH (0.15 henry)

We can't perform any calculations without knowing the signal frequency. We will start by assuming it is 60 Hz. This makes the capacitive reactance equal to:

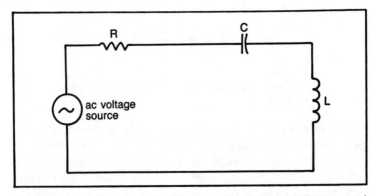

Fig. 2-22. A series resonant circuit is formed by placing an inductance and a capacitance in series.

$$X_c = \frac{1}{6.28 \text{ FC}} = \frac{1}{6.28 \times 60 \times 0.00000022}$$

$$= \frac{1}{0.0000829} = 12057 \text{ ohms} \cong 12000 \text{ ohms} = 12K$$

The inductive reactance works out to:

$$X_L = 6.28FL = 6.28 \times 60 \times 0.15 \cong 56 \text{ ohms}$$

Combining the three resistive values (R, X_c, and X_L), we get a total impedance of:

$$Z = \sqrt{R^2 + (X_L - X_c)^2} = \sqrt{2700^2 + (56 - 12057)^2}$$

$$= \sqrt{7290000 + (-12001)^2}$$

$$= \sqrt{7290000 + 144024000} = \sqrt{151314000}$$

$$= 12301 \text{ ohms}$$

at 60 Hz.

Finally, we can use Ohm's Law to find out how much current is flowing through this circuit at 60 Hz:

$$I = \frac{E}{Z} = \frac{120}{12301} \quad 0.0098 \text{ amp} = 9.8 \text{ mA}$$

Now, let's see what happens when we increase the signal frequency to 250 Hz:

$$X_c = \frac{1}{6.28 \times 250 \times 0.00000022} \quad \frac{1}{0.0003456}$$

$$= 2894 \text{ ohms}$$

$$X_L = 6.28 \times 250 \times 0.15 = 236 \text{ ohms}$$

$$Z = 2700^2 + (236 - 2894)^2$$

$$= \sqrt{7290000 + (2364)^2} = \sqrt{7290000 + 7065532}$$

$$= \sqrt{14355532} = 3789 \text{ ohms}$$

$$I = \frac{120}{3789} = 0.0317 \text{ amp} = 31.7 \text{ mA}$$

Next, we will increase the signal frequency to 1000 Hz:

$$X_c = \frac{1}{6.28 \times 1000 \times 0.00000022} = \frac{1}{0.0013823}$$

$$= 723 \text{ ohms}$$

$$X_L = 6.28 \times 1000 \times 0.15 = 942 \text{ ohms}$$

$$Z = \sqrt{2700^2 + (942 - 723)^2} = \sqrt{7290000 + (219)^2}$$

$$= \sqrt{7290000 + 47981} = \sqrt{7337981} = 2709 \text{ ohms}$$

$$I = \frac{120}{2709} = 0.0443 \text{ amp} = 44.3 \text{ mA}$$

One last example. This time the signal frequency is 5000 Hz:

$$X_c = \frac{1}{6.28 \times 5000 \times 0.00000022} = \frac{1}{0.0069115}$$

$$= 145 \text{ ohms}$$

$$X_L = 6.28 \times 5000 \times 0.15 = 4712 \text{ ohms}$$

$$Z = \sqrt{2700^2 + (4712 - 145)^2} = \sqrt{7290000 + (4567)^2}$$

$$= \sqrt{7290000 + 20863908} = \sqrt{28153908}$$

$$= 5306 \text{ ohms}$$

$$I = \frac{120}{5306} = 0.0226 \text{ amp} = 22.6 \text{ mA}$$

Notice how the impedance starts out high (low current), then decreases as the frequency increases (current flow increases), until a certain point is passed, then the impedance starts to increase (and the current flow decreases) as the signal frequency is increased further. As we shall see in the next section, this crossover point is of considerable significance.

RESONANCE

Since the capacitive reactance decreases as the signal frequency increases, and the inductive reactance increases as the signal frequency increases, at some specific frequency the capacitive reactance will exactly equal the inductive reactance. Well, that only makes sense. So what?

Something very interesting happens at the point where the capacitive reactance equals the inductive reactance. Take a look at how the impedance equation is affected:

$$Z = \sqrt{R^2 + (X_L - X_C)^2} = \sqrt{R^2 + (0)^2}$$

$$= \sqrt{R^2 + 0} = \sqrt{R^2} = R = Z$$

The capacitive and inductive reactance cancel each other out. The ac impedance simply equals the dc resistance. If you think about it for a minute, it becomes clear that this is the minimum value the impedance can ever have.

This condition, where the capacitive and inductive reactances are exactly equal, is called *resonance*. The frequency where this happens is called the *resonant frequency*. There is always one (and only one) resonant frequency for any combination of a capacitance and an inductance.

At resonance, a capacitance and an inductance in series exhibit their minimum impedance. Above or below resonance the impedance will be greater than the dc resistance of the circuit.

How can we determine what the resonant frequency for a given capacitance/inductance combination is? With yet another formula, of course. (I warned you there'd be a lot of math):

$$F = \frac{1}{2 \pi \sqrt{LC}}$$

where F is the frequency in hertz, L is the inductance in henries, and C is the capacitance in farads. Of course, 2π is 6.28 once again.

In the circuit we were dealing with in the last section, the capacitance was 0.22 μF (0.00000022 farad), and the inductance was 150 mH (0.15 henry), so the resonant frequency works out to:

$$F = \frac{1}{6.28 \times \sqrt{0.15 \times 0.00000022}}$$

$$= \frac{1}{6.28 \times \sqrt{0.000000033}} = \frac{1}{6.28 \times 0.001817}$$

$$= \frac{1}{0.0011414} \cong 876 \text{ Hz}$$

Let's check this value out. At 876 Hz, the capacitive reactance is equal to:

$$X_c = \frac{1}{6.28 \times 876 \times 0.00000022} = \frac{1}{0.0012111}$$

$$\cong 826 \text{ ohms}$$

The inductive reactance at 876 Hz works out to:

$$X_L = 6.28 \times 876 \times 0.15 = 826 \text{ ohms}$$

Therefore, the impedance becomes:

$$Z = \sqrt{2700^2 + (876 - 876)^2} = \sqrt{2700^2 + 0}$$

$$= \sqrt{2700^2} = 2700$$
$$Z = R$$

The impedance at 876 Hz. is equal to the dc resistance.

The resonant frequency formula can be rearranged to solve for either of the component values. Notice that the dc resistance has no effect on the resonant frequency.

Let's say we need a circuit that is resonant at 1000 Hz. We will keep our 150 mH coil. What value should we change the capacitor to for the required resonant frequency? First, we rearrange the equation:

$$C = \frac{1}{4 \pi^2 F^2 L}$$

$4 \pi^2$ equals approximately 39.48, so this equation may also be written as:

$$C = \frac{1}{39.48 F^2 L}$$

Plugging in the values for our sample problem, we get:

$$C = \frac{1}{39.48 \times (1000)^2 \times 0.15} = \frac{1}{39.48 \times 1000000 \times 0.15}$$

$$= \frac{1}{5921762.6} = 0.000000169 \text{ farad} = 0.169 \ \mu F$$

Of course, we can check this by returning to the original resonant frequency equation:

$$F = \frac{1}{6.28 \times \sqrt{0.15 \times 0.000000169}}$$

$$= \frac{1}{6.28 \times \sqrt{0.00000002535}} = \frac{1}{6.28 \times 0.0001592}$$

$$= \frac{1}{0.0010004} = 999.6 \text{ Hz.}$$

The small error is due to rounding off values in the calculations. It should be close enough for most practical purposes. After all, many capacitance and inductance values can be off from their nominal value by as much as ± 20%. Clearly a 0.04% error isn't going to matter much.

Similarly, we can rearrange the resonance equation to solve for the inductance:

$$L = \frac{1}{4 \ \pi^2 F^2 C}$$

As an example, let's say we need a circuit that is resonant at 4400 Hz, and we are using a 0.047 μF (0.000000047 farad) capacitor. The required inductance in this case would be:

$$L = \frac{1}{39.48 \times (4400)^2 \times 0.000000047}$$

$$= \frac{1}{39.48 \times 19360000 \times 0.000000047}$$

$$= \frac{1}{35.92} = 0.0278 \text{ henry} = 27.8 \text{ mH}$$

PARALLEL RESONANCE

So far we have been dealing with circuits where the capacitive element and the inductive element are in series, as shown in Fig. 2-22. But what if they are connected in parallel, as illustrated in Fig. 2-23?

For one thing, the impedance equation becomes more complex:

$$z = \sqrt{R^2 \left(\frac{(X_L)(X_C)}{(X_L - X_c)} \right)^2}$$

Let's use the values from our previous example problems:

$$
\begin{aligned}
&\text{ac voltage } = 120 \text{ volts} \\
&\text{R } = 2700 \text{ ohms} \\
&\text{C } = 0.22 \text{ } \mu\text{F } (0.00000022 \text{ farad}) \\
&\text{L } = 150 \text{ mH } (0.15 \text{ henry})
\end{aligned}
$$

There is no need to repeat the reactance calculations for individual frequencies, since they are the same as before:

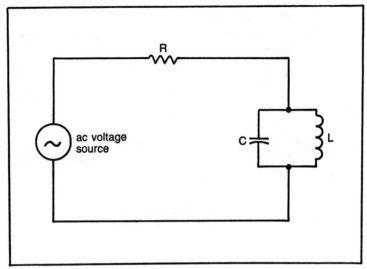

Fig. 2-23. In a parallel resonant circuit, the inductance and the capacitance are in parallel with each other.

60 Hz	$X_c = 12057$	$X_L = 56$
250 Hz	$X_c = 2894$	$X_L = 236$
1000 Hz	$X_c = 723$	$X_L = 942$
5000 Hz	$X_c = 145$	$X_L = 4712$

The impedance at 60 Hz is therefore equal to:

$$Z = \sqrt{2700^2 + \left(\frac{56 \times 12057}{56 - 12057}\right)^2}$$

$$= \sqrt{7290000 + \left(\frac{675192}{-12001}\right)^2}$$

$$\sqrt{7290000 + (-56.26)^2} = \sqrt{7290000 + 3165.3352}$$

$$= \sqrt{7293165.3352} = 2700.5861 \text{ ohms}$$

Raising the signal frequency to 250 Hz changes the impedance to:

$$Z = \sqrt{2700^2 + \left(\frac{236 \times 2894}{236 - 2894}\right)^2}$$

$$= \sqrt{7290000 + \left(\frac{682984}{-2658}\right)^2}$$

$$= \sqrt{7290000 + (-256.95)^2} = \sqrt{7290000 + 66025.41}$$

$$= \sqrt{7356025.41} = 2712.2 \text{ ohms}$$

Now, if the signal frequency is 1000 Hz, the impedance works out to:

$$Z = \sqrt{2700^2 + \left(\frac{942 \times 723}{942 - 723}\right)^2}$$

$$Z = \sqrt{7290000 + \left(\frac{681066}{219}\right)^2}$$

$$= \sqrt{7290000 + (3109.89)^2} = \sqrt{7290000 + 9671418.4}$$

$$= \sqrt{16961418.4} = 4118.42 \text{ ohms}$$

Finally, when the signal frequency is raised to 5000 Hz, the impedance becomes:

$$Z = \sqrt{2700^2 + \left(\frac{4712 \times 145}{4712 - 145}\right)^2}$$

$$= \sqrt{7290000 + \left(\frac{683240}{4567}\right)^2} = \sqrt{7290000 + (149.6)^2}$$

$$= \sqrt{7290000 + 22381.261} = \sqrt{7312381.261}$$

$$= 2704.1415 \text{ ohms}$$

Notice that as the frequency increases, the impedance increases until a specific point is reached, then the impedance starts to decrease with increasing frequency. Once again, the crossover point is the resonant frequency. The resonant frequency is the same for a parallel circuit as for a series circuit, assuming the same components are used in both.

We've already determined that the resonant frequency for this particular combination of components is 876 Hz, when the capacitive reactance and the inductive reactance each equal approximately 826 ohms. Look what happens to the impedance:

$$Z = \sqrt{2700^2 + \left(\frac{826 \times 826}{826 - 826}\right)^2}$$

$$= \sqrt{729000 + \left(\frac{682276}{0}\right)^2} = \sqrt{7290000 + \infty}$$

$$= Z = \infty$$

The impedance is theoretically infinite at resonance in a parallel circuit.

To summarize, in a series circuit, the impedance is at its minimum value (R) at resonance, and in a parallel circuit, the impedance is at its maximum value at resonance.

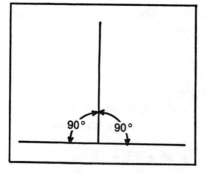

Fig. 2-24. A right angle is 90° from its base line.

TRIGONOMETIC FUNCTIONS

Many electronics calculations require trigonometric functions. Don't panic, it's not quite as complicated as it sounds. It's not easy, but it sounds worse than it is.

Basically, trigonometry is a set of rules for defining relationships between angles and side lengths in triangles. A triangle has only three sides and three angles, so only a limited number of combinations are possible.

We are primarily concerned with right triangles. A right triangle has one right angle. A right angle is a 90° angle. If we have a straight line going form right to left, a line at right angles to it would be going up and down, as illustrated in Fig. 2-24.

A typical right triangle is shown in Fig. 2-25. The longest side (marked c) is called the hypotenuse. It is always directly opposite

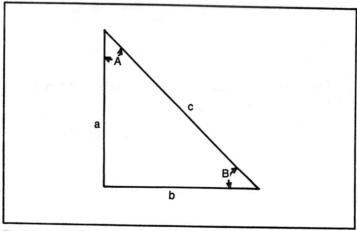

Fig. 2-25. A right triangle includes one right angle. The relationships between the side lengths and the other two angles are the basis of trigonometry.

the right (90°) angle. The length of the sides of a right triangle always bear a specific relationship:

$$c^2 = a^2 + b^2$$

This is true for all right triangles.

Let's say we have a right triangle where side a is 4 inches long, and side b is 5 inches long. How long is side c (the hypotenuse)?:

$$c^2 = 4^2 + 5^2 = 16 + 25 = 41$$

$$c = \sqrt{41} \cong 6.403 \text{ inches}$$

Similarly, there are certain relationships between the angles and the side lengths. There are six basic combinations. Each is given its own name. They are:

sine	cotangent
cosine	secant
tangent	cosecant

For angle A in Fig. 2-25, the formulas are:

$$\text{sine A} = \frac{b}{c} \qquad \text{cotangent A} = \frac{a}{b}$$

$$\text{cosine A} = \frac{a}{c} \qquad \text{secant A} = \frac{c}{a}$$

$$\text{tangent A} = \frac{b}{a} \qquad \text{cosecant A} = \frac{c}{b}$$

The same relationships hold true for angle B, except sides a and b are reversed:

$$\text{sine B} = \frac{a}{c} \qquad \text{cotangent B} = \frac{b}{a}$$

$$\text{cosine B} = \frac{b}{c} \qquad \text{secant B} = \frac{c}{b}$$

$$\text{tangent B} = \frac{a}{b} \qquad \text{cosecant B} = \frac{c}{a}$$

Notice that the B angles can be figured from A, because of the following relationships:

$$\text{sine B} = \text{cosine A} = \frac{a}{c} \qquad \text{cotangent B} = \text{tangent A} = \frac{b}{a}$$

$$\text{cosine B} = \text{sine A} = \frac{b}{c} \qquad \text{secant B} = \text{cosecant A} = \frac{c}{b}$$

$$\text{tangent B} = \text{cotangent A} = \frac{a}{b} \qquad \text{cosecant A} = \text{secant B} = \frac{c}{a}$$

Let's try a few examples, using the triangle from our earlier example. Side a is 4 inches, side b is 5 inches, and side c is approximately 6.4 inches. Therefore:

$$\text{sine A} = \frac{b}{c} = \frac{5}{6.4} = 0.7809$$

$$\text{cosine A} = \frac{a}{c} = \frac{4}{6.4} = 0.6247$$

$$\text{tangent A} = \frac{b}{a} = \frac{5}{4} = 1.2500$$

$$\text{cotangent A} = \frac{a}{b} = \frac{4}{5} \quad 0.8000$$

$$\text{secant A} = \frac{c}{b} = \frac{6.4}{5} = 1.2806$$

$$\text{cosecant A} = \frac{c}{a} = \frac{6.4}{4} = 1.6008$$

In practical electronics work, we will be working from the angle, not side lengths of a hypothetical right triangle. There are mathematical equations for solving the trig functions of an angle, but they are long and complicated. Generally it is more practical to just use a standard trig table.

A table of natural sines from 0° to 90° is given in Table 2-1. Table 2-2 lists cosines, and Table 2-3 lists tangents. These are functions that will normally be encountered in electronics design work.

Cotangents, secants, and cosecants will rarely be used.

Table 2-1. Table to Find Natural Sines.

0 -- 0.0000	25 -- 0.4226	50 -- 0.7660	75 -- 0.9659
1 -- 0.0175	26 -- 0.4384	51 -- 0.7771	76 -- 0.9703
2 -- 0.0349	27 -- 0.4540	52 -- 0.7880	77 -- 0.9744
3 -- 0.0523	28 -- 0.4695	53 -- 0.7986	78 -- 0.9781
4 -- 0.0698	29 -- 0.4848	54 -- 0.8090	79 -- 0.9816
5 -- 0.0872	30 -- 0.5000	55 -- 0.8192	80 -- 0.9848
6 -- 0.1045	31 -- 0.5150	56 -- 0.8290	81 -- 0.9877
7 -- 0.1219	32 -- 0.5299	57 -- 0.8387	82 -- 0.9903
8 -- 0.1392	33 -- 0.5446	58 -- 0.8480	83 -- 0.9925
9 -- 0.1564	34 -- 0.5592	59 -- 0.8572	84 -- 0.9945
10 -- 0.1736	35 -- 0.5736	60 -- 0.8660	85 -- 0.9962
11 -- 0.1908	36 -- 0.5878	61 - -0.8746	86 -- 0.9976
12 -- 0.2079	37 -- 0.6018	62 -- 0.8829	87 -- 0.9986
13 -- 0.2250	38 -- 0.6157	63 -- 0.8910	88 -- 0.9994
14 -- 0.2419	39 -- 0.6293	64 -- 0.8988	89 -- 0.998
15 -- 0.2588	40 -- 0.6428	65 -- 0.9063	90 -- 1.000
16 -- 0.2756	41 -- 0.6561	66 -- 0.9135	
17 -- 0.2924	42 -- 0.6691	67 -- 0.9205	
18 -- 0.3090	43 -- 0.6820	68 -- 0.9272	
19 -- 0.3256	44 -- 0.6947	69 -- 0.9336	
20 -- 0.3420	45 -- 0.7071	70 -- 0.9397	
21 -- 0.3584	46 -- 0.7193	71 -- 0.9455	
22 -- 0.3746	47 -- 0.7314	72 -- 0.9511	
23 -- 0.3907	48 -- 0.7431	73 -- 0.9563	
24 -- 0.4067	49 -- 0.7547	74 -- 0.9613	

Most scientific calculators and computers also include trigonometric functions, so there won't be many occasions when the designer will have to perform the calculations himself. Usually a table, calculator, or computer will be available to do the work for him.

Take a look at the table of natural sines (Table 2-1). Notice how the sine value goes from 0 at 0° to 1 at 90°. Past 90°, up to 180°, the process reverses. The sine decreases from 1 to 0. From 180° to 270° the sine value goes from 0 to -1. From 270° to 360° it goes from -1 back up to 0. Past 360° the sequence repeats and 360° can be considered equivalent to 0°. For values above 360°, subtract 360 as many times as necessary to bring you into the 0° to 360° range.

The sine function through one complete cycle (360° equals one cycle) is graphed in fig. 2-26. Does this graph look familiar? Of course. It is a sine wave, the most basic ac waveform. Now, you know how it got its name.

The cosine function is graphed in Fig. 2-27. Notice that it is the same as the sine function, except shifted by 90°. For the cosine,

Table 2-2. Cosine Values.

0 - - 1.0000	25 - - 0.9063	50 - - 0.6428	75 - - 0.2588
1 - - 0.9998	26 - - 0.8988	51 - - 0.6293	76 - - 0.2419
2 - - 0.9994	27 - - 0.8910	52 - - 0.6157	77 - - 0.2250
3 - - 0.9986	28 - - 0.8829	53 - - 0.6018	78 - - 0.2079
4 - - 0.9976	29 - - 0.8746	54 - - 0.5878	79 - - 0.1908
5 - - 0.9962	30 - - 0.8660	55 - - 0.5736	80 - - 0.1736
6 - - 0.9945	31 - - 0.8572	56 - - 0.5592	81 - - 0.1564
7 - - 0.9925	32 - - 0.8480	57 - - 0.5446	82 - - 0.1392
8 - - 0.9903	33 - - 0.8387	58 - - 0.5299	83 - - 0.1219
9 - - 0.9877	34 - - 0.8290	59 - - 0.5150	84 - - 0.1045
10 - - 0.9848	35 - - 0.8192	60 - - 0.5000	85 - - 0.0872
11 - - 0.9816	36 - - 0.8090	61 - - 0.4848	86 - - 0.0698
12 - - 0.9781	37 - - 0.7986	62 - - 0.4695	87 - - 0.0523
13 - - 0.9744	38 - - 0.7880	63 - - 0.4540	88 - - 0.0349
14 - - 0.9703	39 - - 0.7771	64 - - 0.4384	89 - - 0.0175
15 - - 0.9659	40 - - 0.7660	65 - - 0.4226	90 - - 0.0000
16 - - 0.9613	41 - - 0.7547	66 - - 0.4067	
17 - - 0.9563	42 - - 0.7431	67 - - 0.3907	
18 - - 0.9511	43 - - 0.7314	68 - - 0.3746	
19 - - 0.9455	44 - - 0.7193	69 - - 0.3584	
20 - - 0.9397	45 - - 0.7071	70 - - 0.3420	
21 - - 0.9336	46 - - 0.6947	71 - - 0.3256	
22 - - 0.9272	47 - - 0.6820	72 - - 0.3090	
23 - - 0.9205	48 - - 0.6691	73 - - 0.2924	
24 - - 0.9135	49 - - 0.6561	74 - - 0.2756	

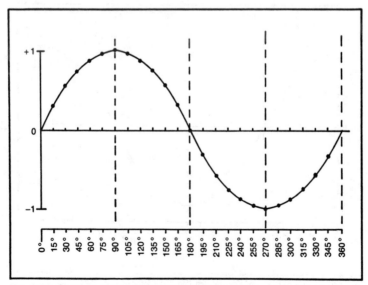

Fig. 2-26. Graphing the sine function gives a smooth, undulating shape.

Table 2-3. Tangent Table.

0 - - 0.0000	25 - - 0.4663	50 - - 1.1918	75 - - 3.7321
1 - - 0.0175	26 - - 0.4877	51 - - 1.2349	76 - - 4.0108
2 - - 0.0349	27 - - 0.5095	52 - - 1.2799	77 - - 4.3315
3 - - 0.0524	28 - - 0.5317	53 - - 1.3270	78 - - 4.7046
4 - - 0.0699	29 - - 0.5543	54 - - 1.3764	79 - - 5.1446
5 - - 0.0875	30 - - 0.5774	55 - - 1.4281	80 - - 5.6713
6 - - 0.1051	31 - - 0.6009	56 - - 1.4826	81 - - 6.3138
7 - - 0.1228	32 - - 0.6249	57 - - 1.5399	82 - - .1154
8 - - 0.1405	33 - 0.6494	58 - - 1.6003	83 - - 8.1443
9 - - 0.1584	34 - - 0.6745	59 - - 1.6643	84 - - 9.5144
10 - - 0.1763	35 - - 0.7002	60 - - 1.7321	85 - -11.43
11 - - 0.1944	36 - - 0.7265	61 - - 1.8040	86 - -14.30
12 - - 0.2126	37 - - 0.7536	62 - - 1.8807	87 - -19.08
13 - - 0.2309	38 - - 0.7813	63 - - 1.9626	88 - -28.64
14 - - 0.2493	39 - - 0.8098	64 - - 2.0503	89 - -57.29
15 - - 0.2679	40 - - 0.8391	65 - - 2.1445	90 - -∞
16 - - 0.2867	41 - - 0.8693	66 - - 2.2460	
17 - - 0.3057	42 - - 0.9004	67 - - 2.3559	
18 - - 0.3249	43 - - 0.9325	68 - - 2.4751	
19 - - 0.3443	44 - - 0.9657	69 - - 2.6051	
20 - - 0.3640	45 - - 1.0000	70 - - 2.7475	
21 - - 0.3839	46 - - 1.0355	71 - - 2.9042	
22 - - 0.4040	47 - - 1.0724	72 - - 3.0777	
23 - - 0.4225	48 - - 1.1106	73 - - 3.2709	
24 - - 0.4452	49 - - 1.1504	74 - - 3.4874	

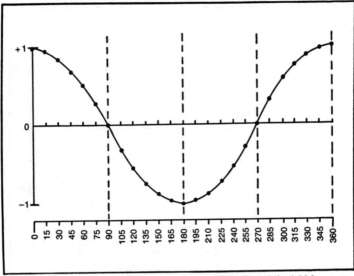

Fig. 2-27. The cosine function is like the sine function shifted 90°.

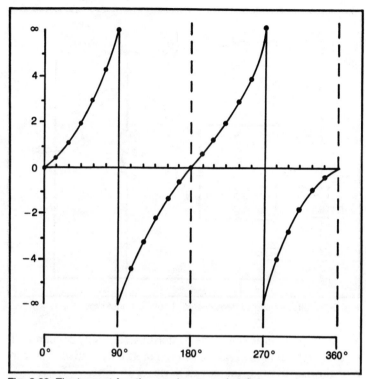

Fig. 2-28. The tangent function reaches towards infinity.

from 0° to 90° the value decreases from 1 to 0. From 90° to 180° it decreases further form 0 to -1. From 180° to 270° it works its way back up from -1 to 0, and from 270° to 360° it climbs from 0 to 1, and the cycle then repeats.

Notice that sine and cosine values are never greater than 1, or less than -1.

Tangents, as listed in Table 2-3, behave somewhat different-ly. From 0° to 90° the tangent value increases from 0 to infinity. Then past 90° to 180° it works its way up from negative infinity to 0. From 180° to 270° we have the exact same pattern as from 0° to 90°, and from 270° to 360° the 90° to 180° pattern is repeated. This is shown graphically in Fig. 2-28.

THE LAPLACE TRANSFORM

Using the mathematical tools described so far we can analyze pretty much what is going on in a circuit if it is being powered by

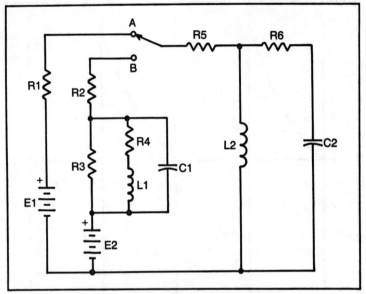

Fig. 2-29. Circuit used to demonstrate Laplace transforms.

a dc voltage, or an ac sine wave. Unfortunately, many circuits we'll be dealing with in the real world are not so cooperative. The current flowing through the circuit varies with time.

Even with a simple dc circuit like the one shown in Fig. 2-29, we sometimes have to be concerned with voltages and currents which change over time. When the switch is in position A, the circuit is effectively as shown in Fig. 2-30. When the switch is moved

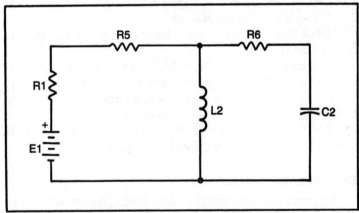

Fig. 2-30. The effective circuit when the switch in Fig. 2-29 is in position A.

to position B, the effective circuit changes to Fig. 2-31. These changes in the circuitry alter the voltages and currents flowing through the components, affecting circuit operation in some way.

Circuits with changing conditions can be solved by using Laplace transforms. Once the correct transforms are found, they can be substituted for the actual circuit values in Kirchhoff equations. The result is then converted into a meaningful value by using a transform table.

There are three major factors to be considered—the circuit elements and their transforms, the initial conditions within the circuit at time 0 (if appropriate), and the time-varying voltages and currents and their transforms.

There are three types of passive components. For a resistor, the Laplace transform is simply the value of that resistor in ohms. The Laplace transform for a 100-ohm resistor is 100.

With inductors and capacitors we start working with a Laplace operator, represented in the equations as s.

The Laplace transform of an inductor is the inductance multiplied by s (sL).

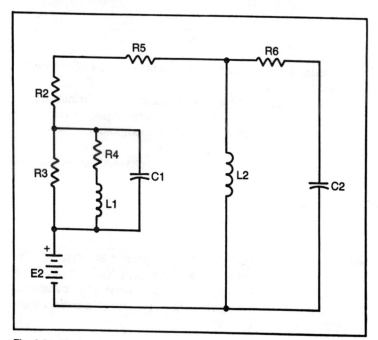

Fig. 2-31. Moving the switch in Fig. 2-29 to position B results in this effective circuit.

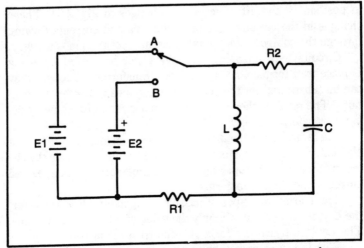

Fig. 2-32. Another circuit to demonstrate the use of Laplace transforms.

The Laplace transform for a capacitor is the reciprocal of s multiplied by the capacitance (in farads). That is:

$$\frac{1}{sC}$$

Consider the circuit shown in Fig. 2-32. At time $t = 0$, the switch is moved from position A to position B. Before $t = 0$, there was a voltage across the capacitor, and a current flowing through the inductor, because there was a complete current path, powered by battery E1. In this case, the initial conditions of the circuit parameters are of importance.

The initial condition transform for a voltage across a capacitor may be expressed as:

$$\frac{E_o}{s}$$

where E_o is the voltage across the capacitor at time $t = 0$. Interestingly, the value of the capacitor is completely irrelevant to the initial condition transform. There is no C value in this transform.

For the current flowing through an inductor, the initial condition transform is written as:

$$LI_o$$

where L is the inductance in henries, and I_o is the current at time $t = 0$. Notice that there is no s expression in this transform.

Both of these transforms may be considered as voltage sources. This concept is illustrated in Fig. 2-33.

A *time function* consists of a coefficient and a function of time. For example, $7e^{-4t}$ is a time function in which the coefficient is 7, and the function of time is e^{-4t}. For another example, $8\sin(5t)$ breaks down to a coefficient of 8, and a function of time equal to $\sin(5t)$.

The Laplace transform of any time-varying function is equal to the coefficient of that function multiplied by the function's transform.

The simplest possible function is 1 (or unit step). This function is applicable in a circuit which initially has no voltage or current flowing through it. At time $t = 0$ a switch is closed, connecting a battery, or other dc voltage source to the circuit. In this case the transform is:

$$\frac{1}{s}$$

Another simple function is the ramp. This is where the voltage starts at zero at time $t = 0$, then increases linearly with time, as illustrated in Fig. 2-34. In this case the transform is:

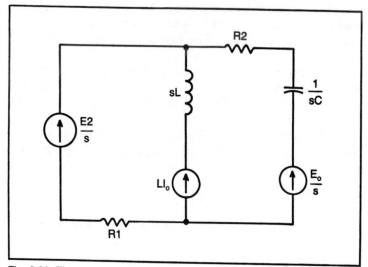

Fig. 2-33. The two transforms may be considered as voltage sources.

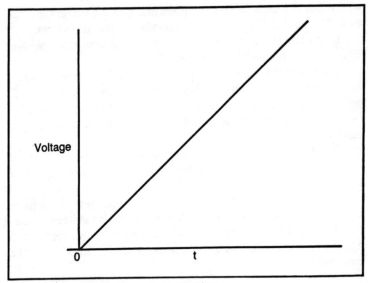

Fig. 2-34. A graph of a linear ramp function.

$$\frac{1}{s^2}$$

Additional common time functions and their transforms are given in Table 2-4. A few of the entries in this table may require a little explanation for many readers.

The e is a mathematical constant, like π. In this case, e is the base of the natural logarithm system (logarithms are discussed elsewhere in this chapter). The approximate value of e is about 2.718.

The term (n-1)! indicates a factorial operation. A factorial is obtained by multiplying the given integer (n-1 in this case) by each lower integer down to 1. That may be a bit confusing, so I will give a couple of examples.

Let's assume n equals 5. That would make n-1 equal to 4, so (n-1)!, or 4! would be equal to the given integer (n - 1, or 4) multiplied by each of the lower integers down to 1. In mathematical terms:

$$4! = 4 \times 3 \times 2 \times 1 = 24$$

Similarly, if n = 10, n -1 will equal 9, so this time (n - 1)! works out to:

$$9! = 9 \times 8 \times 7 \times 6 \times 5 \times 4 \times 3 \times 2 \times 1 = 362880$$

Table 2-4. Laplace Transforms for a Few Common Time Functions.

FUNCTION		TRANSFORM

$$e^{at} = \frac{1}{(s-a)}$$

$$e^{-at} = \frac{1}{(s+a)}$$

$$te^{at} = \frac{1}{(s-a)^2}$$

$$te^{-at} = \frac{1}{(s+a)^2}$$

$$1 - e^{at} = \frac{-a}{s(s-a)}$$

$$\frac{1}{(a-b)}(e^{at} - e^{bt}) = \frac{1}{(s-a)(s-b)}$$

$$\frac{+\dot{e}^{-at}}{(b-a)(c-a)} + \frac{e^{-bt}}{(a-b)(c-b)} +$$

$$\frac{e^{-ct}}{(a-c)(b-c)} = \frac{1}{(s+a)(s+b)(s+c)}$$

$$\frac{t^{(n-1)}e^{at}}{(n-1)!} = \frac{1}{(s-a)^n}$$

$$\frac{\sin(wt)}{w} = \frac{1}{s^2 + w^2}$$

$$\cos(wt) = \frac{s}{s^2 + w^2}$$

$$1 - \cos(wt) = \frac{w^2}{s(s^2 + w^2)}$$

$$e^{-at}\sin(wt) = \frac{w}{(s+a)^2 + w^2}$$

$$\sin(wt + 0) = \frac{(s \times \sin(0)) + (w \times \cos(0))}{s^2 + w^2}$$

$$\cos(wt + 0) = \frac{(s \times \cos(0)) + (w \times \cos(0))}{s^2 w^2}$$

You can see that factorial values can get very large very quickly.

The next term you may find unfamiliar in Table 2-4 is ψ. This term is used to represent radians-per-second. First off, what is *radian*?

The radius of a circle is the length from the center to the edge. The circumference (distance around the outside) of a circle is always equal to $2\pi r$.

Let's say we cut out a segment of a circle, so the portion of the circumference included in the segment is equal to the radius (r). We then draw two radii from the ends of the circumference segment back to the center point of the circle. The two radii will create a specific angle. This angle is equal to one radian. This is illustrated in Fig. 2-35.

A complete circle contains 2π radians.

Radians-per-second is a function of frequency. The formula is:

$$\psi = 2\pi F$$

where F is the frequency in Hertz. The value 2π is approximately equal to 6.28.

The symbol 0 represents a phase angle. If, for instance, a signal is 67° out of phase with a given reference. In this case, 0 would be equal to 67.

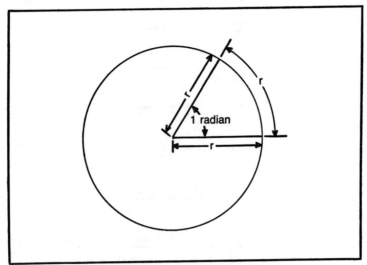

Fig. 2-35. Two radii in a circle define a specific angle.

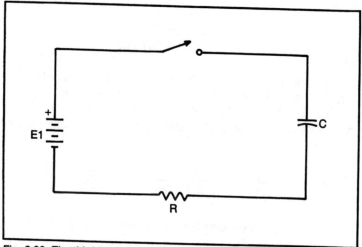

Fig. 2-36. The third Laplace transform demonstration circuit.

While not included in this table, perhaps we should also mention the imaginary *operator j* here. The *j* is equal to the square root of negative one. That is:

$$j = \sqrt{-1}$$

No real number *x* will fulfill the condition:

$$x \times x = -1$$

so *j* is said to be imaginary, but it can have a very real effect in many calculations.

Now we have a table of some of the most commonly used Laplace transforms. What do we do with them?

Once we have the transforms for the circuit elements and have defined the initial conditions and driving functions, we can apply Kirchhoff's Laws, just as we did in the dc circuits earlier in this chapter.

To find the time-varying voltages and currents it is necessary to rearrange the results of the Kirchhoff equations algebraically until they resemble one of the transforms in our table. Then we simply convert to find the results.

As an example of the use of Laplace transforms, let's consider the simple circuit shown in Fig. 2-36. We will assume the following component values:

E1 - 6 volts
R = 2200 ohms (2.2K)
C = 10 μF (0.00001 farad)

The switch is closed at time t = 0. The circuit is redrawn in Fig. 2-37 to show the Laplace transforms of the circuit elements.

The Laplace transform of the resistor is simply equal to the resistance (2200). The Laplace transform for the capacitor works out to:

$$\frac{1}{sC} = \frac{1}{s \times 0.00001} = \frac{100000}{s}$$

The Laplace transform for the voltage source (battery) is:

$$\frac{6}{s}$$

We can combine these transforms into a form analogous to Ohm's Law (E = IR)

$$\frac{6}{s} = \frac{I(2200 + 100000)}{s}$$

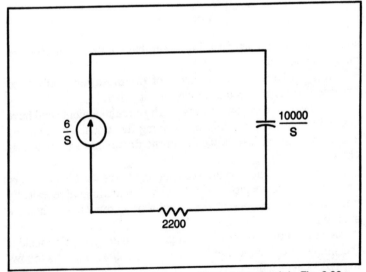

Fig. 2-37. At time t=0, the Laplace transform of the circuit in Fig. 2-36.

Rearranging to solve for I, we get:

$$I = \frac{6/s}{220 = 100000/s} = \frac{6}{2200s + 100000}$$

$$= \frac{0.0027}{s + 45.4545} = 0.0027 \times \frac{1}{s + 45.4545}$$

This closely resembles the second transform in Table 2-4:

$$\frac{1}{(s + a)}$$

0.0027 is the coefficient.

The function for this particular transform is:

$$e^{-at}$$

The variable a holds the same value as before, so the total current flow in this circuit works out to:

$$I = 0.0027 \times e^{-45.4545t}$$

Virtually any time-varying voltage or current in almost any circuit can be determined by using Laplace transforms.

If you feel somewhat confused about Laplace transforms, don't feel too badly. This isn't just math, it's calculus, and there simply isn't enough space here to cover the subject in sufficient depth. A full understanding of Laplace transforms is not absolutely essential for circuit design, but it can be helpful.

LOGARITHMS

Some functions vary smoothly, giving us a nice straight line when we try to graph them. Others tend to have many values jammed together in one part of the scale, while other parts of the scale have values that are widely spaced. The two types of functions are graphed in Fig. 2-38. Figure 2-38A shows a linear function. Figure 2-39B shows a logarithmic function.

Typical logarithmic functions include changes in acoustic volume, and the charging rate of a capacitor.

Logarithmic functions can be treated like any other linear func-

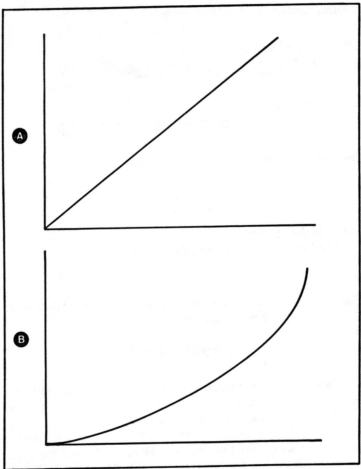

Fig. 2-38. These graphs compare linear and logarithmic functions.

tion, but the wide spread of values can leave you with a lot of numbers which are awkward to work with.

Moreover, logarithms can greatly simplify the math when very large and/or very small numbers are involved.

The old-fashioned slide rule worked on the principle of logarithms.

Before proceeding, we should define just what we mean by a logarithm. The *logarithm* of any specific number is an exponent which indicates the power to which the given base must be raised to equal the given number.

To clarify, let's try a very simple example. Let's assume our logarithmic base is 10. This is often called the common system of logarithms. What if we needed to know the common logarithm of 1000? We want to raise 10 by some unknown power to give us 1000. That is:

$$\log_{10} (1000) = \text{where } 10^x = 1000$$

This particular example is simple enough, since most of us are well aware that 1000 is 10 cubed:

$$1000 = 10 \times 10 \times 10 = 10^3$$

So the common log of 1000 is 3:

$$\log_{10} (1000) = 3$$

Notice how we express the common log of a number N in the form:

$$\log_{10} (N)$$

where the 10 identifies the base.

Solving a common logarithm gets a little more complicated with values that are not exact multiplies of the base. For instance:

$$\log_{10} (657) = 2.8175654$$

Most of us are not used to dealing with noninteger exponents. Most of us would probably be at a loss if we had to determine the value of:

$$10^{2.8175654}$$

It can be done, of course. But the math is fairly complex and tedious. Fortunately, the odds are very, very good that you will never have to perform any such operations yourself. As with the trigonometric functions described earlier, logarithmic tables allow us to bypass most of the busy-work. We can interpolate for values between table entries, or simply round off to the nearest table entry, depending on the level of accuracy required.

Scientific calculators and computers usually include some kind of logarithmic functions too. And, of course, the slide rule can still come in handy for solving logarithmic problems.

Table 2-5. A Few Typical Common Logarithms.

N	$\log_{10}(N)$	N	$\log_{10}(N)$
0	$-\infty$	40	1.6021
1	0.0000	45	1.6532
2	0.3010	50	1.6990
3	0.4771	55	1.7404
4	0.6021	60	1.7781
5	0.6990	65	1.8129
6	0.7782	70	1.8451
7	0.8451	75	1.8751
8	0.9031	80	1.9031
9	0.9542	85	1.9294
10	1.0000	90	1.9542
11	1.0414	95	1.9777
12	1.0792	100	2.0000
13	1.1139	110	2.0414
14	1.1461	120	2.0792
15	1.1761	130	2.1139
16	1.2041	140	2.1461
17	1.2304	150	2.1761
18	1.2553	160	2.2041
19	1.2787	170	2.2304
20	1.3010	180	2.2553
25	1.3979	190	2.2787
30	1.4771	200	2.3010
35	1.5441	300	2.4771
400	2.6021	500000	5.6990
500	2.6990	1000000	6.0000
600	2.7781	10000000	7.0000
700	2.8451	100000000	8.0000
800	2.9031	1000000000	9.0000
900	2.9542	10000000000	10.0000
1000	3.0000		
2000	3.3010		
3000	3.4771		
4000	3.6021		
5000	4.6990		
6000	3.7781		
7000	3.8451		
8000	3.9031		
9000	3.9542		
10000	4.0000		
20000	4.3010		
30000	4.4771		
40000	4.6021		
50000	4.6990		
60000	4.7781		
70000	4.8451		
80000	4.9031		
90000	4.9542		
100000	5.0000		

Table 2-5 lists the common (base 10) logarithms for a number of values.

Another common system of logarithms is the *natural,* or *Napierian system.* This system uses *e* as its base. The *e* is a mathematical constant with a value of approximately 2.718. Why on earth would we want to work with such an odd-ball base? Well, the precise mathematical reasons are quite complex, but basically the answer is that the natural system of logarithms gives us more convenient results for certain types of calculations. It corresponds directly to certain phenomena in nature—that's why it is called the natural system of logarithms.

Any value may be used as a base for logarithms. The base will be identified by the value of B in the expression:

$$\log B(N)$$

Some examples:

$\log_{10}(N)$	base is 10
$\log_e(N)$	base is *e*
$\log_3(N)$	base is 3
$\log_{17.805}(N)$	base is 17.805
$\log_\pi(N)$	base is π (3.14)

For practical electronics work, you will be working almost exclusively with common (base 10) and natural (base *e*) logarithms.

If no base is specified, a base of 10 (common) is assumed. For instance:

$$\log(851)$$

is the same as:

$$\log_{10}(851)$$

If you look over Table 2.5 carefully, you should see certain patterns emerging.

Each logarithm consists of two parts—the *characteristic* and the *mantissa.* The characteristic is the portion of the number to the left of the decimal point, and the mantissa is the portion of the number to the right of the decimal point. Practical log tables generally only give the mantissa, because these values will repeat for different characteristics according to simple rules. A practical log table is given in Table 2-6. This table shows common logarithms. Compare

Table 2-6. A Practical Common Log Table.

1.0	.0000	3.6	.5563	6.2	.7924	8.8	.9445
1.1	.0414	3.7	.5682	6.3	.7993	8.9	.9494
1.2	.0792	3.8	.5798	6.4	.8062	9.0	.9542
1.3	.1139	3.9	.5911	6.5	.8129	9.1	.9590
1.4	.1461	4.0	.6021	6.6	.8195	9.2	.9638
1.5	.1761	4.1	.6128	6.7	.8261	9.3	.9685
1.6	.2041	4.2	.6232	6.8	.8325	9.4	.9731
1.7	.2304	4.3	.6335	6.9	.8388	9.5	.9777
1.8	.2553	4.4	.6435	7.9	.8451	9.6	.9823
1.9	.2787	4.5	.6532	7.1	.8513	9.7	.9868
2.0	.3010	4.6	.6628	7.2	.8573	9.8	.9912
2.1	.3222	4.7	.6721	7.3	.8633	9.9	.9956
2.2	.3424	4.8	.6812	7.4	.8692	10.0	1.0000
2.3	.3617	4.9	.6902	7.5	.8751		
2.4	.3802	5.0	.6990	7.6	.8808		
2.5	.3979	5.1	.7076	7.7	.8865		
2.6	.4150	5.2	.7160	7.8	.8921		
2.7	.4314	5.3	.7243	7.9	.8976		
2.8	.4472	5.4	.7324	8.0	.9031		
2.9	.4624	5.5	.7404	8.1	.9085		
3.0	.4771	5.6	.7482	8.2	.9138		
3.1	.4914	5.7	.7559	8.3	.9191		
3.2	.5051	5.8	.7634	8.4	.9243		
3.3	.5185	5.9	.7709	8.5	.9294		
3.4	.5315	6.0	.7782	8.6	.9345		
3.5	.5441	6.1	.7853	8.7	.9395		

these values with the comparable values in the natural logarithm table (Table 2-7). Natural logs, incidentally, are often written in this form:

$$\ln(A)$$

This is the same as:

$$\log_e (A)$$

Table 2-6 only gives the mantissa portion of the common logarithms. For numbers from 1 to 9.9999, the characteristic will be 0. For numbers less than 1 the characteristic will be negative. For numbers greater than 0, the characteristic will be positive. Typical characteristics are summarized in Table 2-8.

To find the common logarithm for any decimal number, simply find the characteristic for the appropriate range from Table 2-8. Then move the decimal point in the original number until it is between 1 and 10, and find the mantissa in Table 2-6. Combine your characteristic and mantissa, and there you have your common logarithm.

Table 2-7. Natural Log Table.

1.0	0.0000	3.6	1.2809	6.2	1.8245	8.8	2.1748
1.1	0.0953	3.7	1.3083	6.3	1.8405	8.9	2.1861
1.2	0.1823	3.8	1.3350	6.4	1.8563	9.0	2.1972
1.3	0.2624	3.9	1.3610	6.5	1.8718	9.1	2.2083
1.4	0.3365	4.0	1.3863	6.6	1.8871	9.2	2.2192
1.5	0.4055	4.1k	1.4110	6.7	1.9021	9.3	2.2300
1.6	0.4700	4.2	1.4351	6.8	1.9169	9.4	2.2407
1.7	0.5306	4.3	1.4586	6.9	1.9315	9.5	2.2513
1.8	0.5878	4.4	1.4816	7.0	1.9459	9.6	2.2618
1.9	0.6419	4.5	1.5041	7.1	1.9601	9.7	2.2721
2.0	0.6931	4.6	1.5261	7.2	1.9741	9.8	2.2824
2.1	0.7419	4.7	1.5476	7.3	1.9879	9.9	2.2925
2.2	0.7885	4.8	1.5686	7.4	2.0015	10.0	2.3026
2.3	0.8329	4.9	1.5892	7.5	2.0149		
2.4	0.8755	5.0	1.6094	7.6	2.0281		
2.5	0.9163	5.1	1.6292	7.7	2.0412		
2.6	0.9555	5.2	1.6487	7.8	2.0541		
2.7	0.9933	5.3	1.6677	7.9	2.0669		
2.8	1.0296	5.4	1.6864	8.0	2.0794		
2.9	1.0647	5.5	1.7047	8.1	2.0919		
3.0	1.0986	5.6	1.7228	8.2	2.1041		
3.1	1.1314	5.7	1.7405	8.3	2.1163		
3.2	1.1632	5.8	1.7579	8.4	2.1282		
3.3	1.1939	5.9	1.7750	8.5	2.1401		
3.4	1.2238	6.0	1.7918	8.6	2.1518		
3.5	1.2528	6.1	1.8083	8.7	2.1633		

We'll try a few examples.

$$\log (3500)$$

This value is greater than 1000, but less than 10000, so the characteristic is 3. Next, moving the decimal point, we get:

$$3.5$$

Looking up this value in Table 2-6, we find the mantissa should be:

$$.5441$$

Putting it all together, we find:

$$\log (3500) = 3.5441$$

For our next example, we will use:

$$\log (0.00077)$$

The number is between 0.0001 and 0.001, so the characteristic is -4. Moving the decimal point, the value to look up in Table 2-6 becomes:

7.7

which gives us a mantissa of:

.8865

so:

log (0.00077) = -4.8865

Numbers smaller than 1 always have negative logarithms. One more example:

log (415)

The number is in the 100 to 999 range, so the characteristic is 2. But when we move the decimal point we get:

4.15

There is no entry for 4.15 in Table 2-6. What can we do?

The answer is to interpolate between table values. 4.15 is halfway between 4.1 (.6128) and 4.2 (.6232), so its mantissa should

Table 2-8. The Characteristic of a Common Logarithm Indicates the Number of Zeros in the Original Number.

range	characteristic
0.0000000001 to 0.000000000999	-10
0.000000001 to 0.000000009999	-9
0.00000001 to 0.00000009999	-8
0.0000001 to 0.0000009999	-7
0.000001 to 0.000009999	-6
0.00001 to 0.00009999	-5
0.0001 to 0.0009999	-4
0.001 to 0.009999	-3
0.01 to 0.09999	-2
0.1 to 0.9999	-1
1 to 9.999	0
10 to 99.99	1
100 to 999.9	2
1000 to 9999	3
10000 to 99999	4
100000 to 999999	5
1000000 to 9999999	6
10000000 to 99999999	7
100000000 to 999999999	8
1000000000 to 9999999999	9
10000000000 to 99999999999	10

be midway between the two listed values. We will approximate it to be equal to about:

$$0.6180$$

This gives us a final result of:

$$Log\ (415) = 2.6180$$

The common logarithm of any decimal value can be found with relative ease using this method.

ANTILOGARITHMS

The reverse of a logarithm is the antilogarithm. Here we start out with a logarithmic value, and convert it to a decimal number. it is usually written in this form:

$$Antilog\ (N)$$

although you may occasionally see it written like this:

$$Log^{-1}(N)$$

Let's try a few samples. First, we will find the antilogarithm of 0.5682. Once again, we resort to Table 2-6, but we use the second (log) columns to look up our initial value, and read the results in the first (value) columns:

$$Antilog\ (0.5682) = 3.7$$

Since the characteristic of the logarithm value was 0, we are done.

What if the characteristic is not equal to 0? It doesn't complicate matters too much:

$$Antilog\ (C.mmmm) = N \times 10^C$$

where C is the characteristic, mmmm is the mantissa, and N is the value found in the table.

As a practical example, let's consider the antilogarithm of 3.2304. First we take just the mantissa (.2304) and find the appropriate value in the log table. In this case we get 1.7. Then we simply multiply this value by 10 raised to the power of the characteristic:

$$Antilog\ (3.2304) = 1.7 \times 10^3 = 1.7 \times 1000 = 1700$$

Let's try another example, this time with a negative logarithmic value. What is the antilog of -2.7160?

Looking up .7160 in Table 2-6 we get a value of 5.2. The characteristic equals -2, so:

Antilog (-2.7160) = 5.2 × 10^{-2} = 5.2 × 0.01 = 0.052

Sometimes you won't find the exact logarithmic value in the table. In such a case you must interpolate between the two closest table entries. For instance, let's say we have to find the antilogarithm of 0.6556. This mantissa is not included in the log table. The two closest table entries are .6532 and .6628, so the value must be somewhere between 4.5 and 4.6. We could round off to the nearest value. .6556 is closer to .6532, so we could just say:

Antilog (0.6556) ≅ 4.5

If we need a little more accuracy, we can interpolate by approximating the distance between the two values:

.6628 - .6532 = .0096
.6556 - .6532 = .0024

The difference between the value we need and the lower known value (4.5) is about one quarter (25%) of the difference between the known values (4.5 and 4.6), so:

Antilog (0.6556) = 4.525

As with logarithms, any antilogarithm can be solved using this method without much trouble, once you get the hang of it.

COMBINING LOGARITHMS

Now let's put our logarithms and antilogarithms to work, and find out why they are so useful. Let's say we need to find the product of:

450000000 × 7800000000

Numbers in this range can get awkward to work with. It's very easy to make a mistake in the number of zeroes, throwing the result off by a considerable amount.

An interesting mathematical relationship among logarithms provides a more reliable solution:

$$\log (A \times B) = \log(A) + \log(B)$$

To find the product of two numbers, we can add their individual logarithms. This gives the logarithm of the product, so taking the antilogarithm of the result gives us our direct result. This is the way multiplication works on a slide rule.

Let's work through our example:

$$\log (450000000) = 8.6532$$

$$\log (7800000000) = 9.8921$$

$$8.6532 + 9.8921 = 18.5453$$

$$\text{Antilog } (18.5453) \cong 3.51 \times 10^{18}$$

$$= 3500000000000000000$$

This technique also comes in handy with combining very large numbers with very small numbers. For example:

$$63000000 \times 0.000084$$

$$\log(63000000) = 7.79993$$

$$\log (0.000084) = -5.9243$$

$$7.7993 + (-5.9243) = 7.7993 - 5.9243$$

$$= 1.8750$$

$$\text{Antilog}(1.8750) = 7.5 \times 10^1 = 7.5 \times 10 = 75$$

This same kind of thing can work for division too, except we subtract instead of add the logarithmic values:

$$\log \frac{(A)}{B} = \log (A) - \log (B)$$

I doubt that anyone would particularly relish the idea of dividing 9200000 by 310000, but the logarithmic method makes it fairly easy to do:

$$\log (9200000) = 6.9638$$

$$\log (310000) = 5.4914$$

$$6.9638 - 5.4914 = 1.4724$$

$$\text{Antilog } (1.4724) \cong 2.97 \times 10^1 = 2.97 \times 10 = 29.7$$

Here's another example:

$$\frac{0.0038}{0.0015}$$

$$\log (0.0038) = -2.5798$$

$$\log (0.00015) = -3.1761$$

$$-2.5798 - (-3.1761) = -2.5798 + 3.1761 = 0.5963$$

$$\text{Antilog } (0.5963) \cong 3.95$$

Once again, this technique comes in especially handy when combining very large and very small numbers. For instance, trying to solve a problem like:

$$\frac{610000}{0.000033}$$

is just begging for an error in decimal point placement. The logarithmic method eliminates that problem:

$$\log (610000 = 5.7853$$

$$\log (0.000033) = -4.5185$$

$$5.7853 - (-4.5185) = 5.7853 + 4.5185 = 10.3938$$

$$\text{Antilog } (10.3938) = 2.475 \times 10^{10}$$

$$= 2.475 \times 10000000000 = 24750000000$$

Another useful application is in raising a value to a specific power. For example:

$$24^5$$

You could just multiply it out:

$$24 \times 24 \times 24 \times 24 \times 24$$

But that's quite tedious, and it's very easy to make a mistake somewhere along the line. But with logarithms, you can use this approach:

$$\log (A^B) = B \times \log (A)$$

For our example of 24^5, this works out to:

$$5 \times \text{Log} (24) \cong 5 \times 1.3802 = 6.9010$$
$$\text{Antilog} (6.9010) \cong 7.96 \times 10^6 = 7.96 \times 1000000$$
$$= 7960000$$

Actually 24^5 works out to 7962624. The difference is due to rounding off values in the logarithmic method, and interpolating between log table entries. In this case the total error worked out top just slightly over 0.03%, which should be accurate enough for most practical applications.

Extracting roots is particularly tough for standard mathematics. It can be done, of course, but the methods are complex and time consuming. For example, how long would it take you to solve for the fourth root of 22?:

$$\sqrt[4]{22}$$

Logarithms simplify the problem to straightforward division, which can further be reduced to simple subtraction. The logarithmic formula for root extraction is:

$$\log (A \sqrt{B}) \quad \frac{\log (B)}{A}$$

So, for our sample problem:

$$\log (\sqrt[4]{22}) = \frac{\log (22)}{4} = \frac{1.3424}{4}$$

We could go ahead and divide directly, but remember, logarithms allow us to reduce division to subtraction, so:

$$\log \frac{(1.3424)}{4} = \log (1.3424) - \log (4)$$

$$\text{Log} (1.3424) \cong 0.1271$$

$$\text{Log} (4) = 0.6021$$

$$0.1271 - 0.6021 = -0.4750$$

$$\frac{1.3424}{4} = \text{Antilog} (-0.4750) = 0.3350$$

Now, we just need to take the antilog of this value to find the fourth root of 22:

$$\sqrt[4]{22} = \text{Antilog} (0.3350) \cong 2.26$$

According to my scientific calculator, the fourth root of 22 is 2.1657368. The logarithmic method brought us very close.

Of course, if you have a scientific calculator handy with the appropriate functions, you won't need to resort to logarithms to solve such problems, since the functions can be performed directly by the machine.

Many values you will encounter in electronics design work will be in logarithmic form, so you will need to be familiar with logarithms and antilogarithms even if you don't ever use them to solve the types of problems discussed in this section.

DECIBELS

Many phenomena in nature and electronics conform to a logarithmic, rather than a linear scale. This makes comparisons between values somewhat difficult. One solution to this problem was the development of the *decibel* (dB) system.

A decibel is actually one tenth of a *bel* (B), but a bel is too large a unit to be practical for our purposes, so we work only with decibels.

The decibel system is a logarithmic method of comparing two values, powers, voltages, currents, or whatever. Decibels are commonly used in audio equipment, such as amplifiers, because our ears happen to perceive volume logarithmically, rather than linearly. A difference of 6 dB represents an approximate doubling of volume. A 1 dB difference would be virtually inaudible.

The formula for converting two powers to dB form is:

$$dB = 10 \times \text{Log} \left(\frac{P2}{P1} \right)$$

where P1 and P2 are the powers to be compared, and dB is the comparison factor in decibels. If the result is positive, P2 is larger than P1. If the result is negative, P1 is larger than P2.

Advertisements for audio amplifiers frequently emphasize the power ratings of the equipment. you'd think that a 50 watt amplifier would offer a considerable advantage over a 20 watt amplifier, but what is the real difference in decibels?

$$dB = 10 \times \log \frac{(50)}{20} = 10 \times \log (2.5) = 10 \times 0.3979$$

$$= 3.979, \text{ or about 4 dB}$$

There hasn't even been a doubling of acoustic power.

Decibels can also be used to indicate voltage gain (the increase in amplitude from the input signal to the output signal). The formula is:

$$dB = 20 \times \log \frac{(E_o)}{E_i}$$

where E_o is the output voltage, and E_i is the input voltage.

Let's say we have a voltage amplifier where the input signal is 2.0 volts, and the output is 35 volts:

$$dB = 20 \times \log \frac{(35)}{2} = 20 \times \log (17.5) = 20 \times 1.24 = 24.8 \text{ dB}$$

This will also work for attenuation, or negative gain. If the input voltage is 2.0, and the output voltage is 0.5:

$$dB = 20 \times \text{Log} \left(\frac{0.5}{2} \right) = 20 \times \text{Log} (0.25) = 20 \times -0.6021$$

$$= -12.0412 \text{ dB}$$

These calculations are made with the assumption that the input and output impedances are not equal. If this is not the case, the formula should be changed to look like this:

$$dB = 20 \times \text{Log} \left(\frac{E_o \times \sqrt{Z_o}}{E_i \times \sqrt{Z_i}} \right)$$

Let's return to our original example where the input voltage equaled 2 volts, and the output voltage equaled 35 volts. This time we will say the input impedance is 600 ohms, and the output impedance is 10K (10,000 ohms). This makes the gain in decibels equal to:

$$dB = 20 \times Log \left(\frac{35 \times \sqrt{10000}}{2 \times \sqrt{600}} \right) = 20 \times Log \left(\frac{35 \times 100}{2 \times 24.5} \right)$$

$$= 20 \times Log (71.44) = 20 \times 1.8539 = 37.078 \cong 37.1 \ dB$$

Compare this to our original result. Input and output impedances can make a big difference.

The same basic formula is used to make decibel comparisons between input and output currents (current gain):

$$dB = 20 \times Log \ \frac{I_o \times \sqrt{Z_o}}{I_i \times \sqrt{Z_i}}$$

Or, when the input and output impedances are equal:

$$dB = 20 \times Log \ \frac{I_o}{I_i}$$

Notice that the decibel is a comparative, not an absolute value. An expression like, "the amplitude of that sound is 37 dB" is meaningless, unless we know what it is being referenced to. In other words, we need to define the 0 dB point.

Generally, if no reference level is given, a standard reference level of 6 millivolts (0.006 volt) across a 600-ohm impedance is assumed.

SUMMARY

This has been a long and difficult chapter. You probably won't completely grasp everything in a single reading. Use this chapter for reference as you read the rest of this book, and as you work on your own circuit designs. You will probably find that some of these formulas will be used a great deal. Others may rarely, if ever, be used in your work. Usually there are several different ways to approach a design problem. It's your choice which to use.

Chapter 3

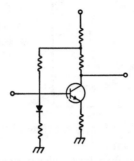

Basic Transistor Circuits

In Chapter 2, we concentrated on passive components (resistors, capacitors, and inductors). Transistors are active components. That is, they are capable of amplification, or signal gain. Therefore, they add a number of new factors to circuit analysis and design.

We discussed the basic construction and operation of transistors in Chapter 1. In this chapter we will explore these devices in more depth and begin to study how they are employed in electronic circuits.

For the most part in the following discussion, we will be dealing with NPN transistors. This is strictly for convenience. PNP transistors work in essentially the same way, except all polarities are reversed, and we are concerned with the flow of holes, rather than the flow of electrons.

In the early years of electronics, current was assumed to flow from the positive terminal of the voltage source to its negative terminal. Today we know that current is actually the flow of electrons from the negative terminal to the positive terminal. Nevertheless, the old positive to negative system is deeply entrenched in electronics. It is often convenient to follow the old tradition of positive to negative conventional current flow.

In understanding the functioning of an electronic circuit, it doesn't really matter whether you use conventional or actual current flow in your circuit analysis, as long as you are consistent in your choice. We will be employing conventional current flow in this

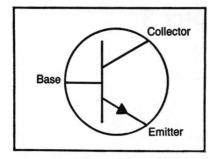

Fig. 3-1. The schematic symbol for an NPN transistor.

chapter, simply because it makes a few of the concepts involved a little easier to explain.

The schematic symbols for bipolar transistors employ conventional current flow. In the symbol for a NPN transistor, as shown in Fig. 3-1, the emitter arrow points out from the base. A handy way to remember this is, "Never Points iN."

For a PNP transistor, on the other hand, the emitter arrow points in towards the base, as illustrated in Fig. 3-2. This can be remembered by thinking, "Points iN Perpetually."

In both cases, the emitter arrow points in the direction of conventional current flow (or, in the opposite direction of actual current flow).

Two semiconductor materials are generally used to manufacture transistors. They are germanium and silicon. Early transistors were primarily germanium, but this has changed over the years. Today silicon transistors are by far the norm. In this book silicon transistors will be assumed, unless otherwise specified. There are some slight differences in the functioning of devices made from these two materials, but the basic principles are the same.

ALPHA AND BETA

The "secret" of the amplification in a bipolar transistor lies in

Fig. 3-2. The schematic symbol for a PNP transistor.

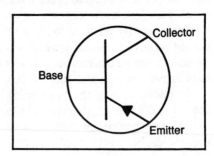

the fact that the collector current is dependent on the base to emitter current.

Figure 3-3 shows the correct polarities for the proper operation of a NPN transistor. This is called *biasing* and will be discussed in more detail later in this chapter.

When biased in this way, more emitter current (I_e) will flow through the transistor than either base current (I_b), or collector current (I_c). In fact, the emitter current is essentially equal to the sum of the other two currents:

$$I_e = I_b + I_c$$

In a practical device, there will be some deviation from this idealized formula due to imperfections in the material, impedance mismatches, stray capacitances, and the like. The differences will be very minor, and can usually be safely ignored. For most practical purposes, the equality will hold true. The designer should be aware of the potential error, however, because it will be of significance in certain applications.

In our discussion we will work with a hypothetical NPN transistor with the following currents flowing through it:

$$I_e = 20 \text{ mA}$$

$$I_b = 0.7 \text{ mA}$$

$$I_c = 19.3 \text{ mA}$$

Notice that these values fit into the formula given above:

$$20 = 0.7 + 19.3$$

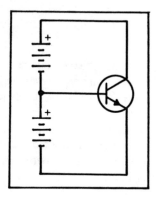

Fig. 3-3. The correct polarities for forward-biasing an NPN transistor.

The ratio of the collector current to the emitter current is called the *alpha* of the transistor. The symbol for alpha is α. The formula for a transistor's alpha is:

$$\alpha = \frac{I_c}{I_e}$$

The alpha is constant for any given transistor. If the emitter current changes, the collector current will change accordingly.

The alpha will always be less than 1. This is because the collector current, by definition, must be smaller than the emitter current:

$$I_c = I_e - I_b$$

Ordinarily, the base current (I_b) will be very small, so the collector current (I_c) will have a value close to that of the emitter current (I_e). This means that normally the alpha of a transistor will be close to, but slightly less than 1.

Let's find the alpha of our hypothetical transistor:

$$\alpha = \frac{I_c}{I_e} = \frac{19.3}{20} = 0.965$$

This is a fairly typical value for alpha.

Another useful relationship among transistor currents is the ratio of the collector current and the base current. This ratio is called *beta*, and its symbol is β. The formula for beta is:

$$\beta = \frac{I_c}{I_b}$$

Since the base current is always very small with respect to the collector and emitter currents, beta will normally have a fairly high value. Typical beta values for silicon transistors range from about 10 to about 1000.

In addition, since we have already determined that the collector current is very close to the emitter current, we can also come up with an approximate value for beta, using this slightly inaccurate formula:

$$\beta \cong \frac{I_e}{I_b}$$

This can come in handy when the collector current is not known.

The value of beta for our hypothetical transistor works out to:

$$\beta = \frac{I_c}{I_b} = \frac{19.3}{0.7} = 27.57$$

Or, using the emitter current in place of the collector current:

$$\beta = \frac{I_e}{I_b} = \frac{20}{0.7} = 28.57$$

Using the emitter current we get a slightly higher value of beta, but it will be quite close to the actual value. The difference won't really be significant in most design calculations.

The value of beta can also be derived from the value of alpha by using this formula:

$$\beta = \frac{\alpha}{\alpha - 1}$$

We have already determined that the alpha for our hypothetical transistor is 0.965, so the beta is:

$$\beta = \frac{0.965}{0.965 - 1} = \frac{0.965}{-0.035} = -27.57$$

We get the same value as from the other beta equation. Don't worry about the negative sign. For our purposes it doesn't mean anything.

Similarly, if we know the value of beta, we can easily find the value of alpha by using this equation:

$$\alpha = \frac{\beta}{\beta + 1}$$

We can check this out with the values for our hypothetical transistor:

$$\alpha = \frac{27.57}{27.57 + 1} = \frac{27.57}{28.57} = 0.965$$

which is the same value we got with our original alpha equation.

Remember, the alpha is always very small (less than 1), while the beta is always fairly large (10 to 1000).

THE COMMON-EMITTER CIRCUIT

There are three basic transistor amplifier circuits. In each of these, one of the transistor's three leads is referenced to the circuit's ground, or common point.

We will start with the *common-emitter* circuit, which is probably the most widely used. In this circuit, the emitter is the common element, just as the name suggests. The emitter is common to both the input signal path, and the output signal path. The input signal is fed into the circuit across the base and the emitter, while the output signal is tapped off across the collector and the emitter.

A simple but typical common-emitter circuit is shown in Fig. 3-4.

Several factors determine the size of the base current (I_b). They include the base supply voltage (E1), the base resistor (R_b), the emitter resistor (R_e), and the voltage drop across the emitter-

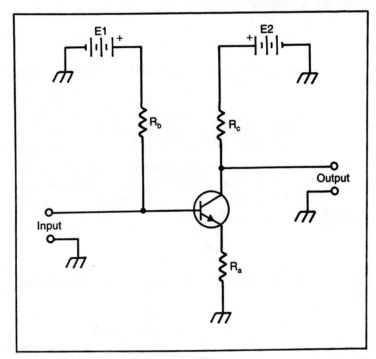

Fig. 3-4. A typical common-emitter circuit.

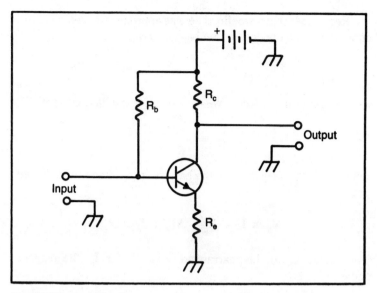

Fig. 3-5. Transistor circuit constructed around a single supply voltage, as shown in this common-emitter circuit.

base junction. This voltage drop is usually about 0.7 volt for silicon resistors. Germanium transistors have a smaller voltage drop across the junction (typically about 0.2 to 0.3 volt).

Most practical designs do not use separate voltage sources for the base and the collector as shown in Fig. 3-4. Instead, a common voltage source is used for both. Resistors R_b and R_c drop the voltage down to the necessary level. This is illustrated in Fig. 3-5.

Since the base current (I_b) flows through resistor R_b, and the emitter current (I_e) flows through resistor R_e, we can calculate the voltage supplied to the base by using this equation:

$$V_b = I_b \times R_b + V_{be} + I_e \times R_e$$

where V_b is the base supply voltage, and V_{be} is the voltage drop across the base/emitter junction. This equation is derived from Ohm's Law:

$$V_b = V_{rb} + V_{be} + V_e$$

$$V_{rb} = I_b \times R_b$$

$$V_e = I_e \times R_e$$

Remember that we defined beta as approximately equal to the emitter current divided by the base current:

$$\beta \cong \frac{I_e}{I_b}$$

Rearranging this equation, we can solve for the base current (I_b):

$$I_B \cong \frac{I_e}{\beta}$$

Substituting this equation in the base supply voltage formula, we get:

$$V_b \cong I_e \times R_b + V_{be} + I_e \times R_e$$

This equation can be rearranged to solve for I_e (the emitter current):

$$V_b - V_{be} \cong I_e \times \frac{R_b}{\beta} + I_e \times R_e$$

$$V_b - V_{be} \cong I_e \times \frac{(R_b + R_e)}{\beta}$$

$$\frac{V_b - V_{be}}{\frac{R_b}{\beta} + R_e} \cong I_e$$

And, since the collector current (I_c) is almost equal to the emitter current:

$$I_c \cong \frac{V_b - V_{be}}{R_b/\beta + R_e}$$

Let's say we are using a silicon transistor with a beta of 100. Since this is a silicon device, the voltage drop across the base/emitter junction is approximately 0.7 volt. In our sample circuit, we will assign a value of 5000 ohms to R_b and 300 ohms to R_e. Plugging these values into the equation, we get:

$$I_c = \frac{V_b - 0.7}{5000/100 + 300} = \frac{V_b - 0.7}{50 + 300} = \frac{V_b - 0.7}{350}$$

In a circuit of this type, the values of V_{be}, R_b, R_e, and β are constants, so the collector current (I_C) is determined by the voltage being fed into the base of the transistor.

In the common-emitter circuit, the base/collector junction is reverse-biased. This means that, theoretically at least, no current should flow through the junction. However, thanks to the minority carriers (see Chapter 1), there is some leakage current (I_{cbo}) across this junction. This is the collector-base current when the emitter circuit is open. When the circuit is complete, it is the current that flows through the base/emitter circuit.

Our approximate beta equation:

$$\beta \cong \frac{I_e}{I_b}$$

can be rearranged to show that the emitter current (I_E) is approximately equal to the base current (I_B) multiplied by the beta (β):

$$I_e = \beta \times I_b$$

If we consider only the leakage currents, the current flowing through the emitter (and the collector) circuit (I_{ceo}) is equal to:

$$I_{ceo} = \beta \times I_{cbo}$$

The total collector current, including the leakage current is:

$$I_c = (\beta \times I_b) + (\beta \times I_{cbo})$$

or:

$$I_c = \beta \times I_b + I_{ceo}$$

In most cases the leakage current will be small enough to be safely ignored in the design equations. In some instances, however, it may throw the equations off enough to affect circuit operation, so the designer should be familiar with the leakage currents.

A common-emitter amplifier has a number of characteristics of interest to the circuit designer.

For one thing, the input impedance is relatively low, while the output impedance is quite high.

Current gain, voltage gain, and power gain can all be large.

The signal is inverted 180° by a common-emitter circuit. That is, when the input goes positive, the output goes negative, and vice

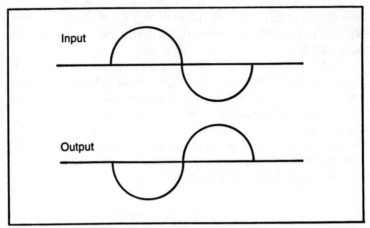

Fig. 3-6. The output of a common-emitter circuit is 180° out of phase with the input.

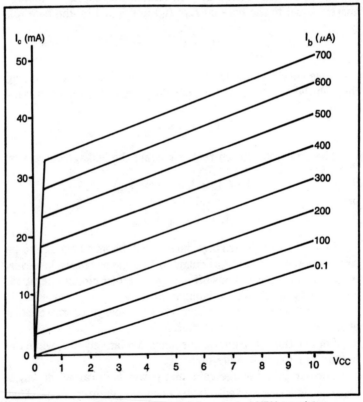

Fig. 3-7. The characteristic curve graph for a typical NPN transistor.

versa, as illustrated in Fig. 3-6. Another way to say this is that the common-emitter circuit inverts any signal fed through it.

Figure 3-7 shows a characteristic curve graph for an NPN transistor in a common-emitter configuration. This graph shows how the collector current (I_c) varies with the collector/emitter voltage (V_{ce}) for different values of base current (I_b).

An example problem will show how this type of graph is used. Let's say we have a base current of 400 μA, and a measured voltage of 5 volts between the collector and the emitter. What is the collector current?

To find the collector current, draw a vertical line from the appropriate point on the V_{ce} line (5 volts) up to where it crosses the 400 μA base current line. From the intersection of these two lines, draw a horizontal line over to the collector current scale. The line will intersect the scale at the appropriate value. This is illustrated

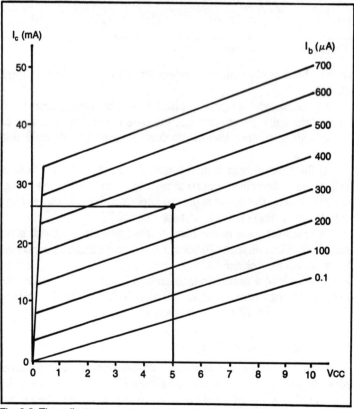

Fig. 3-8. The collector current can be found on the characteristic curve graph.

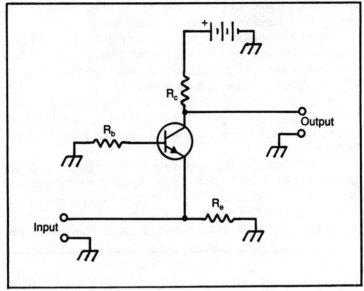

Fig. 3-9. A typical common-base circuit.

in Fig. 3-8. For our sample problem, the collector current is about 28 mA.

Of course, this graph can also be used in the opposite direction. If the collector current and the base current are known, the same basic procedure may be employed to find the collector/emitter voltage.

If the base current is held constant, and the collector voltage is increased, the collector current will also increase. For low values of base current, this effect is not too important, but it becomes more significant at higher levels of base current.

A small increase in the base current will have a much more noticeable effect on the collector current than a comparable increase in the collector voltage.

The *gain* of a basic common-emitter amplifier is equal to the beta of the transistor. That is, the gain is defined as the ratio of the collector current (I_c) to the base current (I_b).

THE COMMON-BASE CIRCUIT

Figure 3-9 shows a typical common-base circuit. Here the common element for both the input and the output is the base. The input signal is fed across the emitter and the base, and the output signal is tapped off across the collector and the base.

104

In this configuration, the signal gain is defined by the ratio of the collector current (I_c) to the emitter current (I_e):

$$\text{Gain} = \frac{I_c}{I_e}$$

Does that equation look familiar to you? It should—it is the formula for the transistor's alpha (α):

$$\alpha = \frac{I_c}{I_e}$$

The gain of a basic common-base amplifier circuit is equal to alpha. Earlier in this chapter we learned that alpha is always less than 1. The gain of a common-base amplifier is always less than unity. That is, the output signal has a lower amplitude than the input signal.

What good is such an amplifier? Attenuation can be achieved with a simpler passive resistance network.

One application is impedance matching, since the input impedance of a common-base amplifier is very low, and the output impedance is very high.

The alpha only expresses the current gain of a common-base amplifier. The voltage gain, on the other hand, can be greater than unity. The power (current times voltage) gain of a common-base amplifier is slightly greater than for a comparable common-emitter amplifier.

The output signal from a common-base circuit is in phase with the input signal. Only the common-emitter configuration inverts the signal.

THE COMMON-COLLECTOR CIRCUIT

The third basic transistor configuration, not surprisingly, uses the collector as the common element. As shown in Fig. 3-10, the input signal is fed across the base and the collector, and the output signal is tapped off across the emitter and the collector.

Notice that the positive terminal of the voltage source is grounded so all operating voltages within the circuit must be negative. The emitter is held at the most negative level.

The voltage gain of a common-collector amplifier circuit is

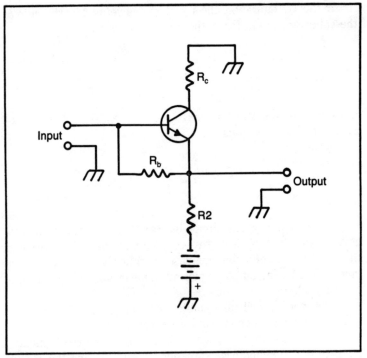

Fig. 3-10. The common-collector circuit.

always less than 1 (below unity). The current gain is fairly high, but the power gain is low.

The common-collector circuit is not used as much as the common-emitter or common-base circuits. Frankly, this configuration simply doesn't make a very good amplifier. The most frequent use of the common-collector circuit is impedance matching. With the common-emitter and common base circuits, the input impedance is lower than the output impedance. If we want to cascade several such amplifier stages in series, we'd have the high impedance outputs of the previous stage(s) feeding the low impedance inputs of the following stage(s). This can lead to inefficient operation, distortion, or even circuit failure.

With the common-collector circuit the impedances are reversed. The input impedance is fairly high, and the output impedance is low. By using a common-collector circuit, a high impedance signal can be converted to a low impedance signal.

The most important characteristics of the three basic transistor circuit configurations are compared in Table 3-1.

Table 3-1. Characteristics of each of the Three Basic Transistor Circuits.

COMMON ELEMENT	EMITTER	BASE	COLLECTOR
INPUT IMPEDANCE	LOW	VERY LOW	MODERATELY HIGH
OUTPUT IMPEDANCE	HIGH	VERY HIGH	LOW
CURRENT GAIN	HIGH	BELOW UNITY	HIGH
VOLTAGE GAIN	HIGH	MEDIUM	BELOW UNITY
POWER GAIN	HIGH	HIGH	LOW
OUTPUT PHASE RELATIONSHIP TO INPUT	180° out of phase	in phase	in phase

BIASING

For proper operation, a transistor must be correctly biased. This simply means that the appropriate polarities and operating voltage levels must be used.

We will consider the correct biasing of a Class A common-emitter amplifier. Amplifier classes will be discussed in more detail in Chapter 4. For now, we will just say that a Class A amplifier is linear over the entire range of input levels. Class A operation is graphed in Fig. 3-11.

The output signal can range from 0 to the supply voltage (V_{cc}). The input signal can go positive or negative. For the no-input condition (input = 0) the output should be half of the full supply voltage. Therefore, the collector should be set at one-half the supply voltage with the input grounded for correct biasing. The no-input condition is often called the *quiescent operating point*.

To complicate biasing somewhat, the one-half the supply voltage rule is actually just an approximate rule of thumb. The ac-

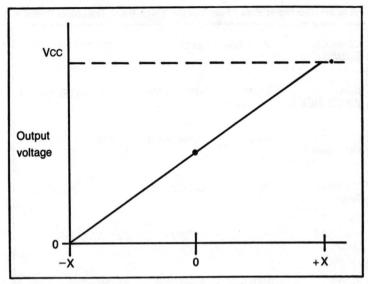

Fig. 3-11. Class A operation is linear throughout its range.

tual operating point values may be different, depending on the individual requirements of the specific circuit and application.

There are a great many different biasing circuits, although they are, for the most part, simply variations of two basic types. In fact, even the two basic types of biasing circuits are essentially variations on each other.

The wide variety of biasing circuits is due to several factors. Device-to-device variation is one factor. Even if two transistors are of the same type number, their actual operating parameters may differ somewhat due to minute variations and inaccuracies that show up in the manufacturing processes. No technology is absolutely perfect.

Operating parameters may also vary with temperature. Beta, for example, is frequently temperature sensitive outside a specific "normal" range. Beta may also fluctuate with changes in the collector current.

The collector current itself is dependent on a number of factors—the collector resistor (R_c), the base supply voltage, the voltage drop across the base/emitter junction (this voltage drop is temperature sensitive too), and the collector to base impedance. If any of these parameters deviates from its nominal value, the collector current could be thrown off. One function of biasing is to correct for such inaccuracies.

108

Biasing also is intended to correct for leakage currents. Early germanium had fairly high collector-base leakage current (I_{cbo}). If the leakage current also flowed across the base/emitter junction (which it normally did in the majority of circuits), it was multiplied by beta, resulting in large leakage currents (I_{ceo}). Moreover, these leakage currents tended to double with each 10 °C temperature increase of the transistor. Biasing circuits were developed to cope with these large, undesirable leakage currents.

Leakage current problems are of somewhat less significance with modern silicon transistors. Often the total leakage current will be small enough to be ignored without affecting circuit operation in any noticeable way. In some circuits, however, the biasing circuitry will have to deal with the leakage currents, even with modern components.

The various bias circuits are intended to stabilize the transistor's operating point. Theoretically, if the proper biasing circuit is used, the operating parameters will not vary with changes in any of the factors mentioned above. Practical circuits don't quite meet this ideal, but they can come close enough for practical applications.

Since biasing is supposed to compensate for variations in circuit parameters, it is vital to understand how a variation in one parameter can cause a change in another parameter. Three *stability factors* are often used in designing bias circuits. These stability factors are identified as S, S_e, and S_β.

The stability factor called S is the ratio of the change in collector current (I_c) and the change in the collector/base leakage current (I_{cbo}):

$$S = \frac{\Delta I_c}{\Delta I_{cbo}}$$

The small triangles (read as "delta") identify the variables as values that change with time.

The second stability factor is S_e. This is the ratio of the change in collector current (I_c) and the change in base voltage (V_{bb}):

$$S_e = \frac{\Delta I_c}{\Delta V_{bb}}$$

Finally, S_β is the stability factor identifying the relationship between a change in beta (β), and a change in the collector current (I_c):

$$S_\beta = \frac{\Delta I_c}{\Delta \beta}$$

Ideally, all three of these stability factors should have values as close to unity as possible. "One" is the perfect stability factor. If two of these stability factors (or all three) have values significantly greater than one, the effects of all variations should be taken into account when designing the bias circuit, and evaluating the operation of the entire circuit as a whole.

The simplest biasing circuit is illustrated in Fig. 3-12. This is a portion of the original common-emitter circuit shown in Fig. 3-4. R_b and V_{bb} set the operating point of the base when no input signal is applied. Similarly, R_c and V_{cc} set the operating point of the collector. This bias circuit assumes that the stability factors are not of significance.

The majority of practical circuits do not use independent voltage supplies for the base and the collector. Instead, a voltage divider network taps the appropriate base bias voltage from the collector voltage source (V_{CC}). Two versions of this approach are illustrated in Fig. 3-13.

All three versions operate in the same manner. We will use

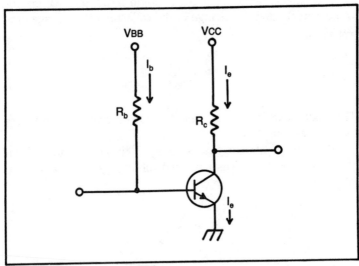

Fig. 3-12. A simple biasing circuit.

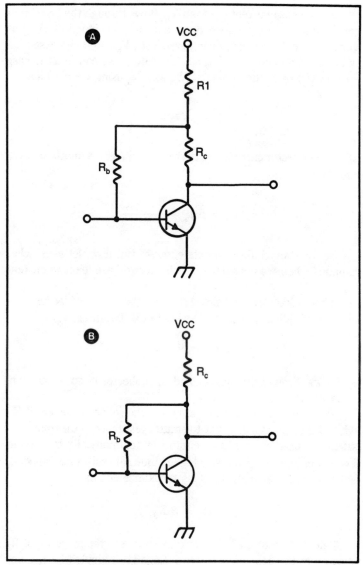

Fig. 3-13. Single supply voltage biasing.

the bias circuit shown in Fig. 3-12 for our discussion, because the separate base and collector voltage supplies will make some of the factors a little clearer. Remember, for the variations shown in Fig. 3-13, the same things are happening, but V_{BB} is defined by the voltage divider network, and originates from V_{CC}.

111

All the base current (I_b) from V_{bb} flows through the base/emitter junction. The voltage drop across that junction (V_{be}) is generally negligible when compared with V_{bb}, so it can essentially be ignored in our calculations. Therefore, we can find the base current (I_b) from the values of V_{bb} and R_b using Ohm's Law:

$$I_b = \frac{V_{bb}}{R_b}$$

The collector current due to the base current is approximately equal to:

$$I_{cb} = \beta\, I_b = \frac{\beta \times V_{bb}}{R_b}$$

So far everything is simple enough. But now, let's see what happens when we consider the collector to base leakage current (I_{cbo}).

This leakage current flows through the base/emitter junction and is multiplied by the value of beta (β), becoming I_{ceo}:

$$I_{ceo} = \beta\, I_{cbo}$$

I_{ceo} flows through the emitter and the collector circuits, substantially affecting their currents.

Another important factor affecting the collector current if R_d, which is the internal collector-to-emitter resistance of the transistor, when it is used in a common-emitter or common-collector circuit. For common-base circuits, a higher internal resistance called R_c is used in the calculations. This resistance equals:

$$R_c = \beta\, R_d$$

The portion of the collector current due to the presence of R_d is equal to the collector/emitter voltage (V_{ce}) divided by R_d:

$$I_{crd} = \frac{V_{ce}}{R_d}$$

The collector/emitter voltage (V_{ce}) is equal to the supply voltage (V_{cc}) minus the voltage drop across collector resistor R_c:

$$V_{ce} = V_{cc} - I_c R_c$$

All these factors can be put together to define the total quiescent (no input signal) collector current (I_c) as:

$$I_c = \beta\, I_b \quad I_{ceo} \quad \frac{V_{cc} - I_c R_c}{R_d}$$

We can rearrange and simplify this equation, so it looks like this:

$$I_c = \frac{\beta\, I_b + I_{ceo} + (V_{cc}/R_d}{R_c/R_d) + 1}$$

This can be further simplified if R_e has a value less than 10% of R_d. In this case, the effect of R_d becomes negligible, and can be eliminated form the equation, leaving us with:

$$I_c = \beta_{I_b} + I_{ceo}$$

Since the base current (I_b) is equal to:

$$I_b = \frac{V_{bb}}{R_b}$$

we can rewrite our formula as:

$$I_c = \frac{\beta\, (V_{bb}\, R_b I_{ceo})}{R_b}$$

This last equation can be used for most designs, where R_d is not of importance. However, R_d can be of considerable importance in calculating ac gain, and output impedances of transistor circuits.

The stability factors can be redefined in constant terms for these circuits. We don't have to bother with the calculus indicated by the changing (delta) values in the original equations. For the circuits shown in Fig. 3-12 and 3-13, the stability factors work out to:

$$S = \beta$$

$$S_e = \frac{\beta}{R_b}$$

$$S_\beta = \frac{I_{ceo} R_b V_{bb}}{R_b}$$

Stability can be improved by adding a resistor (R_e) between the emitter and ground, as shown in Fig. 3-14. In this case the stability factor equations become:

$$S = \frac{\beta (R_e + R_b)}{\beta R_e + R_b}$$

$$S_e = \frac{\beta}{\beta R_e + R_b}$$

$$S = \frac{(R_e + R_b)V_{bb} + I_{cbo} \times R_b \times (R_e + R_b)}{(\beta R_e + R_b) \times (\beta R_e + R_b)}$$

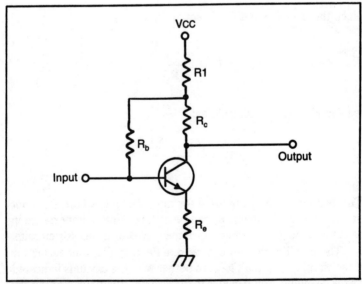

Fig. 3-14. Stability improved by adding a small resistor between the emitter and ground.

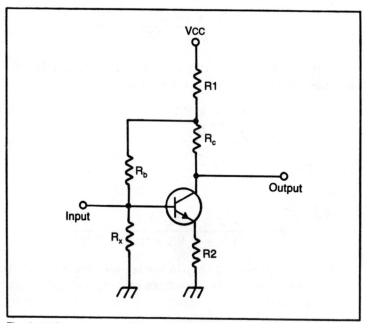

Fig. 3-15. Bias circuit offering increased stability.

The addition of the emitter resistor (R_e) lowers the base current (I_b) because the resistance of R_e is reflected back into the base circuit as a resistance equal to R_e, so the base current becomes:

$$I_b = \frac{V_{bb}}{R_b \; \beta \; R_e} \quad I_{cbo}$$

The collector current becomes equal to:

$$I_c = \beta \; I_b$$

An even more stable bias circuit is illustrated in Fig. 3-15. To simplify the equations, we will define a *combined resistance factor* Rq, which equals:

$$R_q = (R_{eR_c}) + (R_eR_b) + (R_eR_x) + (R_xR_c)$$

The collector current in this circuit takes on a quiescent value of:

$$I_c = \frac{\beta \, (R_x V_{cc} + I_{cbo}(A + R_x R_b))}{\beta \, A + R_x R_b}$$

This bias circuit offers very good stability. The equations for the three stability factors in this circuit are as follows:

$$S = \frac{\beta \, (A + R_x R_b)}{\beta \, A + R_x R_b}$$

$$S_e = \frac{\beta \, R_x}{\beta \, A + R_x R_b}$$

$$S_\beta = \frac{(R_x V_{cc} + I_{cbo} R_x R_b) \times (A + R_x R_b)}{(\beta \, A + R_x R_b) \times (\beta \, A + R_x R_b)}$$

The circuit shown in Fig. 3-16 is often used for temperature compensation. The base/emitter voltage will vary with fluctuations in the temperature of the transistor. This can become especially

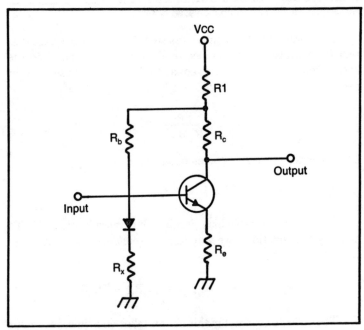

Fig. 3-16. Added circuitry for temperature compensation.

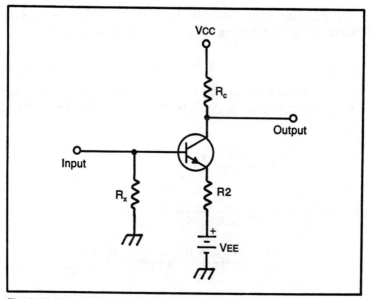

Fig. 3-17. Biasing accomplished with an emitter voltage source.

significant in high power applications like power supplies and large amplifiers where the transistor can heat up considerably.

The diode in this circuit is always conducting. It is selected so its forward voltage drop is equal to the voltage drop across the base/emitter junction of the transistor. The diode and the transistor are positioned as close to each other as possible. This ensures that they will always be at the same temperature, and the voltage across the diode and the base/emitter junction will always be identical. If one goes up due to temperature, the other will also go up by an identical amount. Thus, the voltage across R_e and R_x will always be the same, regardless of any change in V_{be} due to fluctuations in temperature. This circuit offers excellent temperature stability.

We will take a look at one final common biasing circuit before moving on.

In the circuit shown in Fig. 3-17, a small battery (or other voltage source) is inserted between the emitter and ground. This new voltage source is called V_{ee}.

In this arrangement the base current (I_b) is equal to:

$$I_b = \frac{V_{ee}}{R_x + \beta R_e}$$

The collector current, as with the other versions of the bias circuits we've been dealing with is approximately equal to:

$$I_c \cong \beta\, I_b$$

The stability factors for this bias circuit are similar to those for the circuit shown in Fig. 3-14. In this case the stability factor equations are:

$$S = \frac{\beta(R_e + R_x)}{\beta R_e + R_x}$$

$$S_e = \frac{B}{\beta R_e + R_x}$$

$$S_\beta = \frac{(R_e + R_x)V_{ee} + I_{cbo}R_x(R_e + R_x}{(\beta\, R_e + R_x)\,(\beta\, R_e + R_x)}$$

Look back over the biasing circuits shown in Fig. 3-12 through 3-17. Notice how much similarity there is between them. They are all variations on the circuit of Fig. 3-12 or the circuit of Fig. 3-15. Figure 3-15 itself, is essentially a variation of Fig. 3-12.

In designing a biasing circuit, the first step is to decide on the ideal value for the quiescent collector current. Generally, this should be a value at the midpoint between the minimum and the maximum output values. If possible, leave a little"elbow room"on either side of the extremes.

Divide the collector current by the beta to find an approximate value for the base current (I_b). Then design a circuit that will establish those conditions when the input is grounded (no input signal). Which version of the basic bias circuits you start with will depend on the specific application, and the desired stability. Use the simplest circuit that will meet your requirements.

INTRODUCTION TO THE EXPERIMENTS

From here on, each chapter in this book will close with a series of experiments for you to perform. It is strongly recommended that you actually perform these experiments. Hands-on experience is by far the best way to learn. Just reading the text will only get you just so far. To get the maximum value out of this book, get some practical experience with the concepts introduced in the text. The experiments are designed to give you that experience at a minimal expense.

To perform these experiments, you will need some form of breadboarding system. This is simply a method of temporarily hooking up electronic circuits for testing.

Of course, you could hard-wire and solder each test circuit, as if it were intended for permanent use, but this can rapidly become extremely expensive, unless you go to the trouble of desoldering each circuit as you finish with it. This is very, very tedious and time-consuming at best, and you run the risk of damaging some of the more delicate components. Why make more work for yourself? Use a solderless breadboarding system.

Until a few years ago, breadboarding required a bit of ingenuity on the part of the experimenter. Most had to devise their own systems. A few commercial systems were available, but they left quite a bit to be desired in the way of convenience. Tiny spring clips were popular, but a nuisance.

Today, things are much easier. Solderless strip sockets are readily available at low cost. Internally connected rows of holes hold the component leads, and make electrical connections. Virtually any circuit can be breadboarded on such a solderless strip. The holes are spaced for DIP IC packages.

An even better breadboarding system uses a solderless strip socket as the heart of a self-contained package which also includes commonly used subcircuits, such as power supplies and oscillators, and controls such as potentiometers and switches. LED readouts may also be included. You could build your own breadboarding package (it's just a matter of putting all the desired circuits into a single convenient package, with a solderless strip socket on the top), or you can buy a commercially available package.

A good breadboarding package is invaluable for the experimenter—especially if he is designing circuits. the breadboarding package allows the circuit to be tested at various stages of completion. Changes can be easily and quickly made, without risking damage to any of the components.

If you do not have a power supply available, you could use batteries to perform the experiments. The items you will need to perform the experiments in this chapter are outlined in Table 3-2.

EXPERIMENT #1: TESTING A TRANSISTOR WITH AN OHMMETER

In this experiment we will see how a transistor can be check-

Table 3-2. Components and Equipment for the Experiments in Chapter 3.

breadboarding system (solderless strip socket) (see text)
VOM (Volt-Ohm-Milliammeter)
Silicon diode (1N914, or similar)
NPN transistor (2N3904, or similar)
PNP transistor (2N3906, or similar)
+9 volt power supply, or battery
+3 volt power supply, or battery
100 ohm resistor (10 tolerance, 1/2 watt)
1K (1000 ohm) resistor (10% tolerance, 1/2 watt) (2 required)
10K (10,000 ohm) resistor (10% tolerance, 1/2 watt) (2 required)
100K (100,000 ohm) resistor (10% tolerance, 1/2 watt)
470K (470,000 ohm) 10% tolerance, 1/2 watt)
10K potentiometer
hook-up wire

ed for shorts or opens with the ohmmeter section of a standard VOM (Volt-Ohm-Milliammeter).

First we will test a diode. A semiconductor diode has two leads, named the anode and the cathode. The diode is forward biased when the cathode is made negative with respect to the anode.

Most semiconductor diodes have one end marked with a band, or a tapered shape, as illustrated in Fig. 3-18. Take a 1N914 (or similar) diode and connect the end to the negative lead (usually black) of the ohmmeter. The other lead from the diode should be connected to the ohmmeter's positive lead (usually red). This hookup is shown in Fig. 3-19.

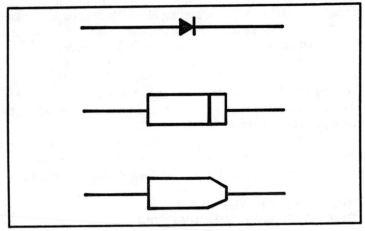

Fig. 3-18. Semiconductor diodes several types of packages.

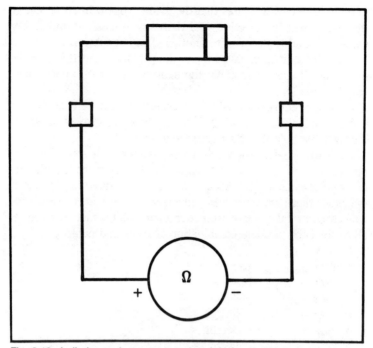

Fig. 3-19. A diode conducts current when it is forward-biased.

Now read the resistance indicated on the ohmmeter. I got approximately 4500 ohms (4.5K). You may get a somewhat different value, but the resistance should be fairly low (less than 10,000 ohms, or 10K). With this hookup, the diode is forward-biased.

If you get a very high, or infinite resistance reading, the diode is probably open. If the resistance reading is extremely low (near zero), the diode could be shorted. Either of these conditions indicate internal damage to the diode, and it should be thrown away.

Assuming you got a good forward-biasing resistance reading on your diode, the next step is to reverse the polarity of the ohmmeter, as shown in Fig. 3-20. This arrangement allows you to measure the reverse-biased resistance of the diode.

This time the ohmmeter's pointer should barely budge from the full scale, infinity position.

As discussed in Chapter 1, a diode has a relatively low internal resistance when it is forward-biased, but a very high internal resistance when it is reverse-biased.

If you got a fairly low resistance reading in both directions, you can be sure the diode is shorted. A very large (or infinity) reading in both directions indicates that the diode is open. In either case, the component is defective, and therefore useless. It should be discarded, and replaced with another diode of the same type.

A 1N914 diode is called for in the parts list, but this experiment should give essentially the same results with almost any standard diode.

The procedure outlined in this experiment can be used to test a diode of unknown quality. This method can also be used to discover the polarity of an unmarked diode.

The same basic procedure can be used for transistors. For our purposes, a transistor can be thought of as the functional equivalent of two diodes connected back to back, as illustrated in Fig. 3-21.

Since there are three leads on a transistor (emitter, base, and collector), a total of six resistance measurements must be made to allow for each possible combination of leads and polarity:

emitter +	base −
emitter −	base +
emitter +	collector −
emitter +	collector +
base +	collector −
base −	collector +

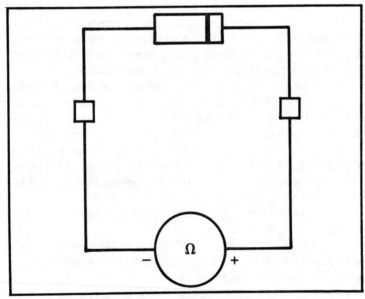

Fig. 3-20. A diode blocks current when it is reverse-biased.

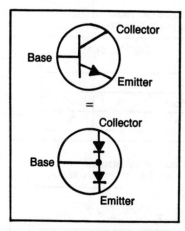

Fig. 3-21. A transistor is almost like two back-to-back diodes.

Make each of these measurements for a 2N3904 NPN transistor (or similar), and write down the readings in the appropriate spaces in Table 3-3.

My results are shown in Table 3-4. While you may have come up with slightly different values for some of the measurements, your results should be similar to mine, and have the same high/low relationships:

emitter +	base −	low
emitter −	base +	high
emitter +	collector −	high
emitter −	collector +	high
base +	collector −	high
base −	collector +	low

Now repeat the six measurements with a PNP transistor (2N3906, or similar). Enter your results in Table 3-5, and compare them with my results in Table 3-6.

Table 3-3. Worksheet to Record Results in Measuring the Internal Resistances of an NPN Transistor.

		POSITIVE LEAD		
		E	B	C
	E	X		
NEGATIVE LEAD	B		X	
	C			X

123

**Table 3-4. Compare Results in Table 3-3
with the Author's Results Shown Here.**

		POSITIVE LEAD		
		E	B	C
NEGATIVE LEAD	E	X	∞	∞
	B	5500	X	5400
	C	∞	∞	X

**Table 3-5. Measure the Internal Resistances
of a PNP Transistor and Record Your Results in this Table.**

		POSITIVE LEAD		
		E	B	C
NEGATIVE LEAD	E	X		
	B		X	
	C		X	

Table 3-6. Compare Your Results in Table 3-5 with My Results Shown Here.

		POSITIVE LEAD		
		E	B	C
NEGATIVE LEAD	E	X	6000	∞
	B	∞	X	∞
	C	∞	5500	X

Now compare the results for the NPN transistor (Tables 3-3 and 3-4) with the results for the PNP transistor (Tables 3-5 and 3-6). Notice how they are essentially mirror images of each other. The measurement value should be approximately the same, but with the polarities reversed.

If you did not use 2N3904 and 2N3906 transistors, you may see somewhat more variation in the measurements for the two transistors. The 2N3906 is more or less a PNP version of the 2N3904.

The basic mirror image pattern should hold true for any NPN and PNP transistor pair, however.

EXPERIMENT #2: COMMON-BASE AMPLIFIER

In this experiment we will start working with the basic amplifier

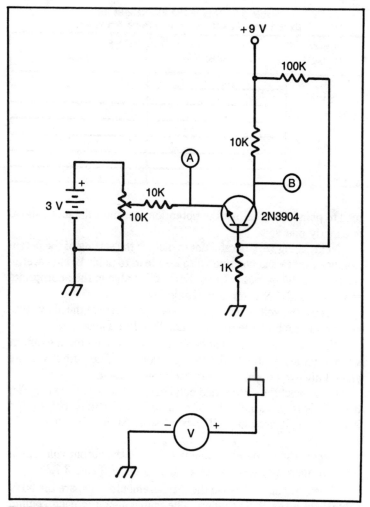

Fig. 3-22. The circuit for the first part of experiment #2—the NPN common-base amplifier.

circuits. Carefully breadboard the circuit shown in Fig. 3-22. This is a common-base circuit built around an NPN transistor. Double check all the connections before applying power to the circuit.

Notice that one lead (the positive lead) of the voltmeter is left unconnected in the schematic diagram. If this lead is connected to point A, the meter will display the input voltage. If the free lead is moved to point B, the output voltage will be indicated.

For the first step in this experiment, connect the free voltmeter

Table 3-7. This is the Worksheet for
Experiment #2—the Common-Base Amplifier.

INPUT VOLTAGE	OUTPUT VOLTAGE	
	NPN	PNP
0 volt		
0.25 volt		
0.5 volt		
1 volt		
1.5 volt		
2 volts		
2.5 volts		

lead to point A and adjust the potentiometer until the input signal is exactly one volt.

Next, being very careful not to disturb the setting of the potentiometer, move the positive voltmeter lead to point B and carefully measure the output voltage. Enter this value in the appropriate space in the NPN column of Table 3-7.

Return the voltmeter lead to point A, and repeat the above procedure for each of the input values listed in Table 3-7.

Now, examine the table carefully, and notice how the output voltage varies in step with the input voltage. The output voltage should always be greater than the input voltage.

Disconnect the circuit, and build the PNP version shown in Fig. 3-23. This is essentially the same amplifier circuit as before, but it is now built around a PNP transistor. Notice the changes in polarity.

Repeat the entire experiment and enter the output voltages in the appropriate spaces in the PNP column of Table 3-7.

When you have finished the measurements, compare the NPN column with the PNP column. The values listed in each column should be very close for each specific input voltage. There may be some minor variations due to component tolerances.

EXPERIMENT #3: COMMON-EMITTER AMPLIFIER

Repeat the procedure of Experiment #2 with the common-emitter amplifier circuit shown in Fig. 3-24. Enter your results in the worksheet provided in Table 3-8.

Only the NPN transistor version is used in this experiment, since we already proved in the last experiment that a PNP transistor circuit would give essentially the same results, once all the

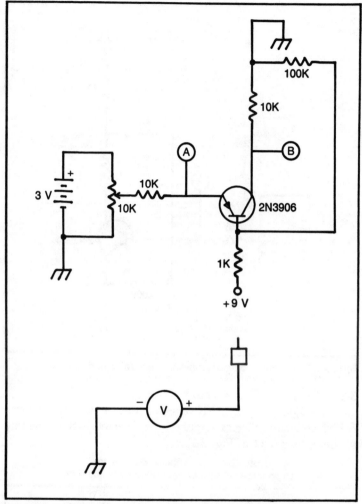

Fig. 3-23. The PNP common-base amplifier is like the NPN version, but with the polarities reversed.

circuit polarities are reversed. As an additional exercise, you might want to rearrange the circuit shown in Fig. 3-24 for use with a PNP transistor, such as the 2N3906.

EXPERIMENT #4: COMMON-COLLECTOR AMPLIFIER

Repeat the procedure outlined in Experiment #2 once more, this time using the common-collector amplifier circuit illustrated in Fig.

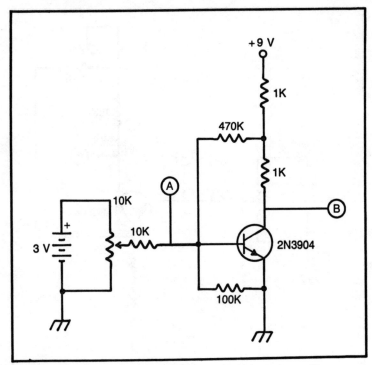

Fig. 3-24. The common-emitter amplifier circuit for experiment #3.

3-25. Use Table 3-9 to record your results.

Once again, only the NPN version of this circuit is shown here. A PNP common-collector amplifier would behave in the same way as the NPN circuit of Fig. 3-25.

Table 3-8. Use This Worksheet for Experiment #—the Common-Emitter Amplifier.

INPUT VOLTAGE	OUTPUT VOLTAGE
0 volt	
0.25 volt	
0.5 volt	
0.75 volt	
1 volt	
1.5 volt	
2 volts	
2.5 volts	

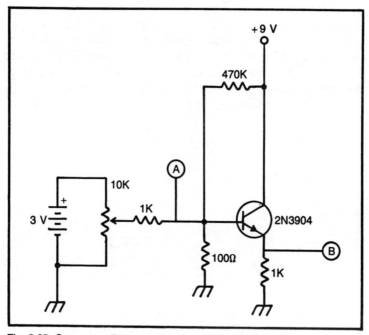

Fig. 3-25. Common-collector circuit for experiment #3.

SUMMARY

Now you should have a fairly good idea of what a transistor is, and how it functions. You should also be familiar with a number of equations that can be used to analyze electronic circuits.

With this material firmly under your belt, you are at last ready

Table 3-9. Record your Results in This Worksheet for Experiment #4—the Common-Collector Amplifier.

INPUT VOLTAGE	OUTPUT VOLTAGE
0 volt	
0.25 volt	
0.5 volt	
0.75 volt	
1 volt	
1.5 volt	
2 volts	
2.5 volts	

to begin working on the design of actual practical circuits. It would probably be a good idea to go back and review the first three chapters of this book before moving on to Chapter 4. A lot of material has been presented so far. You certainly don't have to memorize all the equations. Even professional designers have to look up a formula from time to time. But you should be able to recognize and understand the formulas when you see them.

Chapter 4

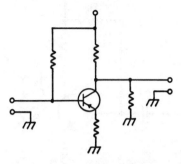

Amplifiers

The most basic application for a transistor is in an amplifier circuit, which increases the amplitude of an input signal. The output signal is a larger replica of the input signal. (Actually, some amplifier circuits decrease the signal level, and should, more properly be called attenuators, but the principles of operation are the same.)

The three basic transistor circuits (common-emitter, common-base, and common-collector) discussed in the preceding chapter are essentially amplifier stages, although they can be put to work in other applications (as we will see in later chapters).

In this chapter we will study the ways transistors can be used in practical amplifier circuits.

AC AMPLIFIERS

The simple basic amplifier circuits presented in Chapter 3 function as described only with dc input signals. If the input signal fluctuates over time (that is, if it is an ac signal) a number of additional factors must be considered to create a functional amplifier circuit. Some parameters (such as beta (β)) may change from the nominal dc value when ac signals are used. Stability can become a problem. Additional components are needed for a functional ac amplifier circuit. These include a load resistor, and capacitors to prevent the ac signal from disturbing the dc bias voltages needed for proper transistor operation.

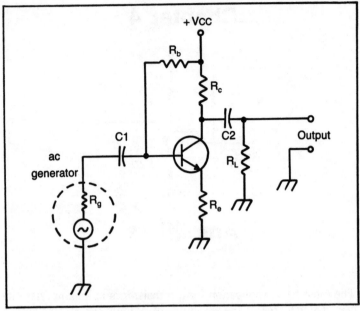

Fig. 4-1. A simple common-emitter ac amplifier.

Figure 4-1 shows a simple common-emitter ac-amplifier circuit. Resistance R_g represents the internal impedance of the input signal source itself. Notice the overall similarity between this circuit and the basic common-emitter circuits presented in the last chapter. Besides the ac input signal source, three components are added to the basic common-emitter circuit. They are capacitors C1 and C2, and load resistor R_L.

Capacitor C1 blocks the base bias current provided by R_b from flowing back into the signal source and flowing through R_g to ground. If that happened, the quiescent operating point of the circuit would be upset. Capacitor C1's value is fairly large, so the ac input signal can pass through to the base of the transistor for amplification.

The amplified output signal appears across collector resistor R_c, and is fed to the output and load resistor R_L through capacitor C2, which blocks the dc collector voltage from reaching the output. The amplified ac signal, however, passes through C2 to appear across load resistor, R_L.

This also works in the opposite direction. If capacitor C2 was not in the circuit, the ac output signal could throw off the dc bias voltage applied to the collector.

According to Ohm's law, the voltage drop across the collector resistor R_c is:

$$V_{R_c} = R_c \times I_c$$

where I_c is the dc collector current.

Obviously this means the dc voltage reaching the collector must be equal to:

$$V_C = \text{VCC} - V_{R_c} = \text{VCC} - (R_c \times I_c)$$

Now, if capacitor C2 was not in the circuit, a complete dc current path from VCC through resistors R_c and R_L to ground would exist. An additional current (I_L) due to R_L would be added to I_c to make up the total dc current flowing through R_c. The voltage across this resistor would become:

$$V_{R_c} = R_c \times (I_c + I_L)$$

This would cause the voltage at the collector (V_c) to drop to a value of:

$$V_c = \text{VCC} - (R_c \times (I_c + I_L))$$

No matter how small I_L may be, our new value of V_c would have to be lower than that determined by the earlier version of the equation. Capacitor C2 prevents this from happening.

Resistor R_L represents the ac load on the output of the transistor. The output is the voltage across this resistor. R_L may not be an actual resistor. It may represent the impedance of a loudspeaker, or the input impedance of an additional amplifier stage, or other circuit.

In our discussion of the basic common-emitter circuit in Chapter 3, we learned that the output current is beta (β) times the input current. A similar relationship exists with the ac amplifier, but here we must work with the ac beta, which may or may not be equal to the dc beta.

In technical literature, ac beta is sometimes referred to as h_{fe}. More commonly, however, just the Greek letter β is used. To prevent confusion between ac beta and dc beta, the dc value is usually written as β_{dc}. We will employ this convention throughout the rest of this book. No subscript is used for ac beta, because it is the value normally used in design equations, so it is convenient to just write β, and ac beta will be assumed.

The equation for ac beta is very similar to the one for dc beta:

$$\beta = \frac{\Delta I_c}{\Delta I_b}$$

The small triangles (read as "delta") indicate a changing value. In other words, ac beta is the ratio between the change in collector current (I_c) and the change in base current (I_b).

One way to find beta is with a set of characteristics curves, as shown in Fig. 4-2. The quiescent operating voltage (V_{ceq}) is selected, and the common point for the desired base current (I_{b3} in our example) and the appropriate collector (I_{c3} in our example) is found. this is marked as point Q in the diagram.

Since dc beta is just the straightforward ratio of these two currents, the value can be readily found:

$$\beta_{dc} = \frac{I_c}{I_b}$$

For ac beta, we are concerned with the ratio of changing currents, so the equation becomes somewhat more complicated. We

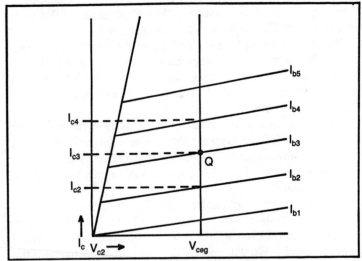

Fig. 4-2. The characteristic curves graph for a transistor.

can find the value for ac beta by selecting additional crossing points on either side of Q. The equation then becomes:

$$\beta = \frac{\Delta I_c}{I_b} = \frac{I_{c4} - I_{c2}}{I_{b4} - I_{b2}}$$

By the same token, ac alpha is the ratio between the change in collector current and the change in emitter current:

$$\alpha = \frac{\Delta I_c}{\Delta I_b}$$

(From here on, dc alpha will be written as α_{dc}. If just α is used, ac alpha will be assumed.)

It generally isn't practical to try to determine ac alpha from the characteristic curves, like ac beta. This is because the value is, by definition, very close to unity (1). Fortunately, the relationship between alpha and beta we learned about for dc circuits still holds. Therefore, once we know the ac beta, we can readily solve for the ac alpha, using this formula:

$$\alpha = \frac{\beta}{1 + \beta}$$

Of course, this is the same equation we used with the dc values.

When working with an ac amplifier, we also have to consider some differences in the emitter resistance. For a dc circuit, this was just a discrete resistor (R_e) connected between the transistor's emitter and ground. This resistor is the only resistance of significance in the dc circuit. But when an ac signal is being amplified by the transistor, the internal resistance of the emitter itself becomes important. This resistance is identified in the equations as r_e. The small r indicates it is an internal resistance, rather than an independent resistor. It is not too difficult to find the value for the internal emitter resistance, using this simple formula:

$$r_e = \frac{26}{I_c}$$

where I_c is the collector current in milliamperes, and r_e is the internal emitter resistance in ohms.

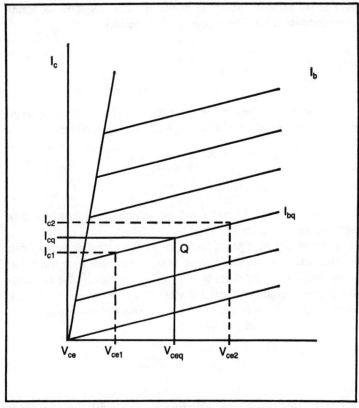

Fig. 4-3. The characteristic curves graph used to determine the collector current.

The total resistance in the emitter circuit is simply the sum of the internal emitter resistance, and the external emitter resistor:

$$R_{et} = R_e + r_e$$

As with the dc circuit, the emitter resistance (R_{et}) can be reflected back into the base circuit, where its value is multiplied by the ac beta.

The internal collector resistance is dependent on the circuit configuration used. In a common-base circuit, r_c is the internal collector-to-base resistance of the transistor. For common-collector, and common-emitter circuits, the internal collector-to-base resistance is called r_d.

The easiest way to find the value of r_d is to use a set of characteristic curves, as shown in Fig. 4-3. Once the quiescent

136

operating voltage (V_{ceq}) is determined, select two additional voltages that are equidistant, and on opposite sides of the quiescent value. These additional voltages are labeled V_{ce1} and V_{ce2} in the diagram. Find the collector current (I_{c1} and I_{c2}) for these two voltages, assuming that the base current (I_b) remains at its quiescent level. The value of the internal collector resistance (r_d) can now be found using these derived values:

$$r_d = \frac{V_{ce2} - V_{ce1}}{I_{c2} - I_{c1}}$$

As you have no doubt guessed, there is also an internal base resistance in a transistor. However, this resistance is usually quite small (typically between 500 and 1000 ohms), and can reasonably be ignored in the majority of design equations.

As discussed in Chapter 3, a common-emitter circuit features a low input impedance and a high output impedance. This is true for ac circuits, as well as dc circuits.

For the simple circuit shown in Fig. 4-1, the input impedance seen by the signal source (generator) works out to:

$$R_i = \frac{R_b \times (r_b + \beta (R_e + r_e))}{R_b + R_b + \beta (R_e + r_e)}$$

Since the value of r_b is generally insignificant, we can simplify this equation to:

$$R_i = \frac{R_b \times \beta (R_e + r_e)}{R_b + \beta (R_e + r_e)}$$

The output impedance seen looking back into the collector circuit is:

$$R_o = \frac{r_d((R_g + r_b) + \beta (R_e + r_e))}{R_g + r_b + R_e + r_e}$$

Notice that the internal impedance of the signal source (R_g) and the base resistance (r_b) are reflected from the input of the transistor to its output.

Once again, the r_b factor can usually be ignored, simplifying the equation to:

$$R_o = \frac{r_d(R_g + \beta\,(R_e\,r_e))}{R_g + R_e + r_e}$$

This formula gives us the actual output impedance of the transistor itself. The output impedance seen by the load resistance (R_L) is this value in parallel with the collector resistor (R_c). That is:

$$R_{oL} = \frac{R_o \times R_c}{R_o + R_c}$$

The current gain in the ac circuit (A_i) is simply the ac beta. That is:

$$A_i = \beta$$

The formula for the ac voltage gain (A_v) is somewhat more complex:

$$A_v = \frac{R_L \times R_c}{R_L + R_c} \times \frac{1}{R_e + r_e}$$

The ac power gain (G), of course, is simply the product of the current gain and the voltage gain:

$$G = A_i \times A_v$$

We now are ready to try our hand at actually designing a common-emitter amplifier circuit.

A TYPICAL DESIGN

In our design, we will be using the circuit shown in Fig. 4-1.

The first step in any design is to determine what characteristics we want the circuit to have.

Let's say we want our circuit to run from a 9-volt battery. This gives VCC a value of 9 volts.

In our sample design, we will assume the load we want to drive has an input impedance of 10K (10,000 ohms). This defines the value for R_L.

The most important preliminary design we need to make is the amount of gain. In our sample design, we will aim for a voltage gain of 15.

The next step is to choose a transistor and find its ac beta. We'll assume the manufacturer's specification sheet for the transistor we decide to use states that the ac beta is 250.

Now we are ready to get to work on the actual design of the circuit. The first step is to select a value for the collector resistor (R_c). This resistor should have a value that is between 10% and 25% of that of the load resistor (R_L). For our circuit, R_c must therefore be between:

$$R_{cmin} = 10\% \text{ of } 10000 = 0.1 \times 10000 = 1000 \text{ ohms}$$

and:

$$R_{cmax} = 25\% \text{ of } 10000 = 0.25 \times 1000 = 2500 \text{ ohms}$$

By keeping R_c within this range, the ac load line will not greatly differ from the dc load line.

For our design, we will select a value of 2200 ohms for R_c. This is a standard value within the desired range.

The total external resistance within the collector circuit is the parallel combination of R_c and R_L:

$$R_{CC} = \frac{2200 \times 10000}{2200 + 10000} = \frac{22000000}{12200} \cong 1800 \text{ ohms}$$

For the desired voltage gain of 15, the ratio of the resistance in the collector circuit to the resistance in the emitter circuit should be 15:

$$15 = \frac{R_{cc}}{R_{ee}} = \frac{1800}{R_{ee}}$$

Rearranging this equation, we get:

$$R_{ee} = \frac{1800}{15} = 120 \text{ ohms}$$

The total resistance in the emitter circuit should therefore be

120 ohms. This value is made up of the internal emitter resistance (r_e) and the external emitter resistor (R_e)

$$R_{ee} = R_e + r_e = 120$$

Clearly, we cannot select a value for the emitter resistor (R_e) until we know the value of r_e. To find that value, we first need to know the collector current (I_c).

Our power supply VCC has already been defined as 9 volts. Ideally the quiescent collector voltage should be half of that (4.5 volts). This would allow the output to swing from 0 to 9 volts equally around the 4.5 quiescent point.

To achieve this, 4.5 volts must be across the collector resistor, which has a value of 2200 ohms in our circuit. Therefore, according to Ohm's Law:

$$I_c = \frac{V_c}{R_c} = \frac{4.5}{2200} = 0.002 \text{ amps} = 2 \text{ mA}$$

We can now use this value to solve for r_e, using the equation:

$$r_e = \frac{26}{I_c} = \frac{26}{2} = 13 \text{ ohms}$$

Now, we can select a value for the emitter resistor (R_e). For the desired gain of 15, it may be as high as:

$$R_e = R_{ee} - r_e = 120 - 13 = 107 \text{ ohms}$$

We can meet the desired conditions with a standard 100 ohm resistor for R_e.

To get the desired collector current (I_c) of 2 mA, the base current (I_b) must be 2 mA, divided by the beta of the transistor. That is:

$$I_b = \frac{I_c}{\beta} = \frac{2}{250} = 0.008 \text{ mA} = 8 \text{ }\mu A$$

In the circuit of Fig. 4-1, base current flows through resistor R_b, originating from the 9-volt VCC supply. The base current also flows through the base-emitter junction, and through R_e to ground.

This means the total resistance I_b flows through is equal to:

$$R_{bt} = R_b + r_e + R_e$$

Using Ohm's Law, we can find a value for this total resistance:

$$R_{bt} = \frac{V_{CC}}{I_b} = \frac{9}{0.000008} = 1125000 \text{ ohms}$$

Since we already know $R_e = 100$ and $r_e = 13$, it is a simple matter to find the necessary value for R_b:

$$\begin{aligned} R_b = R_{bt} - R_e - r_e &= 1125000 - 100 - 13 \\ &= 1124887 \text{ ohms} \end{aligned}$$

We can round this off and use a 1.2 megohm (1200000 ohms) resistor for R_b.

The values for the two capacitors (C1 and C2) are not particularly critical, except in a few specialized frequency sensitive applications. Generally any value from about 0.01 μF to 0.1 μF can be used.

We have now designed our common-emitter amplifier. The parts values are:

$$\begin{aligned}
R_b &= 1.2 \text{ megohm (1200000 ohms)} \\
R_c &= 2.2K \quad (2200 \text{ ohms}) \\
R_e &= 100 \text{ ohms} \\
R_L &= 10K \ (10000 \text{ ohms}) \\
C1 &= 0.01 \ \mu F \\
C2 &= 0.01 \ \mu F
\end{aligned}$$

To see how well we did, we first need to know the source impedance (R_g). Let's say it is 10,000 ohms. This resistance behaves as a voltage divider with the input impedance of the circuit. We can ignore the minimal effect of r_b in calculating the input impedances:

$$R_i = \frac{R_b \times \beta \ (R_e + r_e)}{R_b + \beta \ (R_e + r_e)} = \frac{1200000 \times 250 \times (100 + 13)}{1200000 + 250 \times (100 + 13)}$$

$$= \frac{1200000 \times 250 \times 113}{1200000 + 250 \times 113} = \frac{1200000 \times 28250}{1200000 + 28250}$$

$$= \frac{33900000000}{0.0000008} = 27600 \text{ ohms}$$

Because of the voltage divider effect, only part of the original input signal actually reaches the transistor for amplification. The actual input signal, as seen by the base of the transistor works out to:

$$\frac{R_i}{R_i + R_g} \times V_{in} = \frac{27600}{27600 + 1000} \times V_{in} = \frac{27600}{37600} = 0.73 V_{in}$$

Only 73% of the input signal is amplified. The overall voltage gain of the circuit is the voltage gain of the transistor multiplied by 0.73.

The voltage gain of the transistor, you should recall, can be found with this formula:

$$A_v = \frac{R_L \times R_c}{R_L + R_c} \times \frac{1}{R_e + r_e}$$

$$= \frac{10000 \times 2200}{10000 + 2200} \times \frac{1}{100 + 13} = \frac{22000000}{12200} \times \frac{1}{113}$$

$$= 1803.28 \times 0.00885 = 15.96$$

The actual gain is only 73% of this value, so the gain of this circuit is:

$$A_v = 15.96 \times 0.73 = 11.65$$

This is somewhat lower than our desired nominal value of 15. To compensate for the effect of the voltage divider at the input, it is necessary to decrease the value of emitter resistor R_e.

Let's try another design example. We will assume the following starting conditions:

$$
\begin{array}{rcl}
V_{CC} & = & 12 \text{ volts} \\
R_L & = & 15000 \text{ ohms (15K)} \\
A_v & = & 30 \text{ (desired value)} \\
R_g & = & 25000 \text{ ohms (25K)} \\
\beta & = & 300
\end{array}
$$

The collector resistor (R_c) may be anywhere from 1500 ohms to 3750 ohms. We will use a standard 3300 ohm (3.3K) resistor for R_c.

The total resistance in the collector circuit is the parallel combination of R_c and R_L:

$$R_{cc} = \frac{3300 \times 15000}{3300 + 15000}$$

$$= \frac{49500000}{18300} \cong 2705 \text{ ohms}$$

This makes the maximum resistance for the emitter circuit equal to:

$$R_{ee} = \frac{R_{cc}}{A_v} = \frac{2705}{30} = 90 \text{ ohms}$$

Notice how the emitter resistance is always a relatively small value.

Assuming we are using the ideal quiescent collector voltage (6 volts, for a VCC of 12 volts), the collector current a value of:

$$I_c = \frac{6}{3300} = 0.0018 \text{ amp} = 1.8 \text{ mA}$$

This gives us a value for r_e of:

$$r_e = \frac{26}{1.8} = 14.4 \text{ ohms}$$

so the maximum allowable value for R_e is:

$$R_e = 90 - 14.4 = 75.6 \text{ ohms}$$

We will use a 68 ohm resistor.

The next step is to find the base current (I_b):

$$I_b = \frac{I_c}{\beta} = \frac{1.8}{300} = 0.006 \text{ mA} = 6 \text{ } \mu\text{A}$$

The total base circuit resistance (R_{bt}) equals:

$$R_{bt} = \frac{Vcc}{I_b} = \frac{12}{0.000006} = 2000000 \text{ ohms}$$

The values of R_e and r_e are so small, that they can be ignored, and we can use a 2 megohm (2000000 ohms) resistor for R_b.

The input impedance for this circuit is:

$$R_i = \frac{R_b \times (R_e + r_e)}{R_b + (R_e + r_e)} = \frac{2000000 \times 300 \times (68 + 14.4)}{2000000 + 300 \times (68 + 14.4)}$$

$$= \frac{2000000 \times 300 \times 82.4}{2000000 + 300 \times 82.4} = \frac{2000000 \times 24720}{2000000 + 24720}$$

$$= \frac{49440000000}{2024720} = 24418 \text{ ohms}$$

Due to the voltage divider effect, the input voltage is attenuated by an amount equal to:

$$\frac{24418}{24418 + 25000} = \frac{24418}{49418} = 0.494 = 49.4\%$$

As you can see, the input voltage drop can be quite severe in some circuits.

The nominal gain of the circuit is:

$$A_v = \frac{15000 \times 3300}{15000 + 3300} \times \frac{1}{68 + 14.4}$$

$$= \frac{4950000}{18300} \times \frac{1}{82.4} = 2704.92 \times 0.01213$$

$$= 32.83$$

Allowing for the voltage divider effect, this becomes:

$$A_v = 32.83 \times 0.494 = 16.22$$

If we need a gain of 30, a gain of 16.22 certainly wouldn't be satisfactory.

If we change the value of R_e, we don't need to go back and redo all the other calculations. We can approximate the altered gain. Let's try dropping the value of R_e to 33 ohms. This makes the transistor gain equal to:

$$A_v = 2704.92 \times \frac{1}{33 + 14.4}$$

$$= 2704.92 \times \frac{1}{47.4}$$

$$= 2704.92 \times 0.0211 = 57.07$$

This is considerably higher than our desired gain of 30. But when we multiply it by the voltage divider factor, it drops to:

$$A_v = 57.07 \times 0.494 = 28.19$$

That's still slightly lower than our nominal gain of 30, but it should be close enough for most practical applications. If it is not close enough, R_e would have to be lowered further.

Care must be taken that the transistor gain does not exceed the specifications of the transistor used. Excessive current and/or voltage can destroy the delicate semiconductor crystal. There is a limit to how much gain you can get from a transistor.

CASCADING AMPLIFIER STAGES

A transistor is not a miracle device. There is a definite limit to how much a single transistor can amplify a signal. For very large gains, you have to divide the labor between two or more transistor amplifier stages.

Sequentially combining amplifier stages is called *cascading*. Very large gains can be achieved by cascading amplifier stages. As an example, let's say we have a three-stage amplifier with the following individual voltage gains for each stage:

A_{v1}	30
A_{v2}	20
A_{v3}	25

The total overall gain of the amplifier as a whole is the product of the individual stage gains:

$$A_{vt} = A_{v1} \times A_{v2} \times A_{v3} = 30 \times 20 \times 25 = 15000!$$

Clearly, cascading amplifier stages permits gains that would be utterly impossible for a single transistor.

With a gain of the size used in our example, different transistors would be used in each of the stages. The first stage would probably be built around a small, low-power transistor. This stage will most likely only be dealing with input and output signals in the milliwatt range. The final stage, on the other hand, would have to be a heavy-duty power transistor. If a 5 mV signal is fed to the input of our sample amplifier, the output produced by the third transistor after the gain of 15000 would be 75 volts. That would take one heck of a transistor!

The input and output impedances of each stage are important when cascading amplifier stages. The most efficient power transfer will occur when the source impedance equals (or is close to) its loaded impedance. Sometimes, however, a deliberate small impedance mismatch will be used to reduce distortion, but in most cases, source and load impedances should be as close as possible.

Impedance matching is complicated by the fact that impedance, by definition, varies with frequency. The source may have a nominal impedance of 10000 ohms at 1 khz, and the load impedance may nominally be 10000 ohms at 1 khz too. But at 500 khz, the source impedance may be 8000 ohms, and the load impedance 23000 ohms. If the reactance of the source is primarily inductive, and the reactance of the load is primarily capacitive, this could occur. (Remember, for an inductor, reactance increases as frequency increases, and for a capacitor, reactance decreases as frequency increases.)

The input and output impedances for the three basic types of transistor amplifier circuits are summarized in Table 4-1.

There are nine possible ways to cascade a pair of the three amplifier stage types.

SOURCE	LOAD
common-base	common-base
common-base	common-emitter
common-base	common-collector

SOURCE

common-emitter
common-emitter
common-emitter
common-collector
common-collector
common-collector

LOAD

common-base
common-emitter
common-collector
common-base
common-emitter
common-collector

Each of these combinations are rated in Table 4-2. Three of the combinations are poor, or extremely poor, and should be avoided.

common-base to common-base
common-emitter to common-base
common-collector to common-collector

In a few noncritical applications you may be able to get away with a common-base to common-emitter connection, but the performance of the combination will leave a lot to be desired. It would probably be best to avoid this combination too.

A common-emitter to common-emitter connection will usually give acceptable performance, but it's certainly not the best choice.

Four of the nine possible combinations offer good performance:

common-base to common-collector
common-emitter to common-collector
common-collector to common-base
common-collector to common-emitter

Notice that all the good cascading combinations involve a common-collector stage. Unfortunately, a common-collector stage always has a negative voltage gain, and a low power gain (although it can

**Table 4-1. The Three Basic Amplifier
Circuits Have Different Input and Output Impedances.**

Circuit Type	Input Impedance	Output
common-base common-emitter common-collector	very low low medium-high	very high high low

**Table 4-2. Some Combinations of the BASIC
Amplifier Circuits Make Between Impedance Matches Than Others.**

Source	Match to Load		
	common-base	common-emitter	common-collector
common-base	1	3	5
common-emitter	2	4	5
common-collector	5	5	1

1	extremely poor
2	poor
3	mediocre
4	acceptable
5	good

offer high current gain). Using a common-collector stage can often defeat the whole purpose of cascading amplifier stages.

The common-emitter to common-emitter combination is probably the most useful for cascading to increase voltage or power gain, even though it does not have the best impedance matching characteristics.

Another factor to be considered when working with common-emitter stages is phase relationships. A common-emitter circuit inverts the input signal, shifting it $180°$. In many applications this won't matter, but in some cases it can be important. A second common-emitter stage will re-invert the signal, so the final output will be back in phase with the original input. An even number of common-emitter stages will have an output that is in phase with the input. An odd number of common-emitter stages will result in an output that is $180°$ out of phase with the input. Common-base and common-collector stages do not invert the phase of the signal.

This discussion of cascading amplifier stages is assuming that direct connections are used. Of course, the impedance mismatches can be compensated for by inserting an impedance matching stage between amplifier stages. Impedance matching transformers are frequently used, but they can have a damaging effect on the frequency response of the amplifier as a whole. A common-collector stage with near unity gain can be used for impedance matching in many circuits.

The obvious way to directly couple a pair of amplifier stages is illustrated in Fig. 4-4. Unfortunately this usually will not work, because the dc bias voltages of the second stage are thrown way

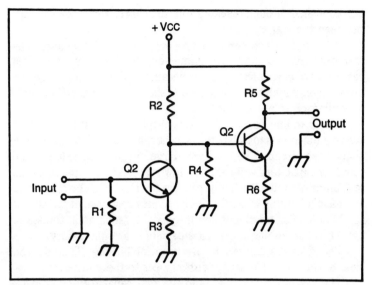

Fig. 4-4. Direct coupling is not always the best way to cascade multiple amplifier stages.

off by the first stage. The base of the second transistor has too high a positive voltage on it. It may be equal too, or even greater than the collector voltage of the second transistor. Clearly, the transistor cannot function because it is not correctly biased. The full collector voltage of the first transistor is fed to the base of the second transistor.

In some cases the circuit resistances can be adjusted to allow for correct biasing of the second transistor, but this is tricky, at best.

Another common solution is to use complementary stages. That is, if the first stage is built around an NPN transistor, the second stage is built around a PNP transistor. A fairly typical circuit of this type is shown in Fig. 4-5. This is called *complementary-symmetry direct coupling*.

Q1 is an NPN transistor. Q2 is a PNP complement of Q1. The two transistors have identical (or nearly identical) characteristics, except for their operating polarities. Because the various specifications (alpha, beta, collector voltage, emitter voltage, etc.) are the same for both transistors, the stages are said to be symmetrical.

Both stages are common-emitter circuits, and are correctly biased for normal operation.

Another approach to cascading amplifier stages is to use RC coupling. As the name suggests, the coupling between stages is accomplished through resistors and capacitors, to provide isolation between the stages.

RC coupling can compensate somewhat for impedance mismatches between stages. Common-emitter to common-emitter cascades are particularly popular because of their high voltage and power gain capabilities. Figure 4-6 shows a typical RC coupled cascading of amplifier stages.

The coupling capacitors are typically around 0.1 μF. Their primary function is to isolate the dc collector voltage from the subsequent base circuit. The relatively large capacitance value of the coupling capacitors will exhibit a low reactance, compared to the resistance of their associated resistors (marked R). The capacitors will have a limiting effect on the frequency response of the circuit as a whole. For relatively low frequencies (such as in the audio range), relatively large capacitance values are mandatory.

The resistors marked R_b are used for biasing. They set the base to the correct polarity with respect to the emitter.

The resistor marked R_d and the two capacitors marked C_d make up a decoupling network. Their function is to prevent undesirable interaction and feedback between stages through the common impedance of the Vcc voltage source.

Three main factors limit the frequency response of this circuit.

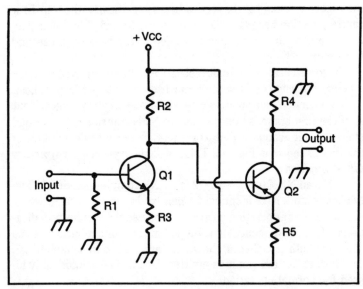

Fig. 4-5. One method of cascading amplifier stages is to use complimentary stages.

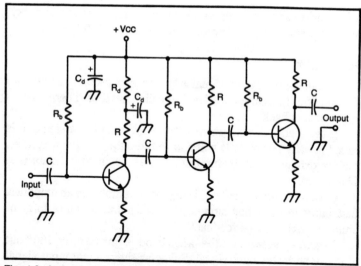

Fig. 4-6. A simple RC network used to cascade multiple amplifier stages.

The size of the coupling capacitors has already been mentioned. The larger these capacitors are, the lower the useful frequencies can be.

The other two factors are due primarily to the internal construction of the transistor itself. Each stage exhibits a specific input and output capacitance. These capacitances, which are due mainly to internal capacitances within the transistor, tend to shunt high frequencies to ground. For good high frequency response, these capacitances clearly should be as low as possible. Transistors suitable for high frequency (radio) work have very low internal capacitances.

The final major factor limiting frequency response is also due to the internal construction of the transistor. This is the *transit time*, which is the length of time it takes the electrons to move from one lead to another. Some modern transistors feature very low transit times, and therefore, have very good high frequency response.

Coupling stages with mismatched impedances can also be accomplished with interstage transformers, as shown in Fig. 4-7. The turns winding ratio in the transformer is selected to match the output impedance of stage 1 and the input impedance of stage 2. The capacitor is used to isolate the dc VCC voltage. While once very common, transformer coupling isn't used very often these days, except in very low-end equipment. This is because of poor frequency response, and distortion introduced by the inductance of the amplifier. In addition, the coupling transformers are rather bulky, and the trend today is towards very compact circuitry.

FEEDBACK

If some of the output signal is looped back around and fed into the amplifier for reamplification, it is called *feedback*. There are two types of feedback.

If the feedback is in phase with the original input signal, it is called *positive feedback*. It will add to the output of the amplifier, rapidly driving it into saturation. It may begin to oscillate (see Chapter 5).

Clearly, positive feedback is highly undesirable in an amplifier. Instability in amplifier circuits is usually due to some form of unintentional positive feedback.

Negative feedback, on the other hand is inverted, or 180° out of phase with the input signal. Since, as the original input signal is going up, the feedback signal is going down, and vice versa, part of the signal reaching the amplifier is canceled out. Negative feedback reduces the overall gain of the circuit. Why on earth would we want to do that? There are actually many reasons. The most common include improved stability, lower distortion, and scaling of the output signal to match a specific range.

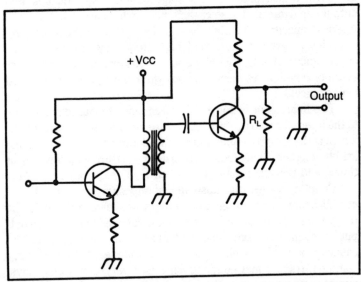

Fig. 4-7. Multiple amplifier stages cascaded by using impedance matching transformers.

What happens if the feedback signal is out of phase with the input signal, but not by 180°. For instance, the feedback signal may be 47° or 225° out of phase. Portions of the signal will cancel each other out (acting as negative feedback), while other portions of the signal will reinforce each other (acting like positive feedback). A feedback signal with any phase other than 0° (in phase) or 180° (inverted) is very difficult to work with, and seldom desirable.

Some confusion can arise when talking about feedback in amplifier circuits. The ratio of feedback signal to total output signal:

$$\frac{V_f}{V_o}$$

is traditionally identified as β. This is a throwback to vacuum tube days. Because of the potential for confusion with the beta of a transistor (the I_c/I_b ratio), we will avoid using the beta symbol for the *feedback ratio*. We will use B for this purpose. Be very careful not to get the two symbols confused. Be especially careful when dealing with other technical literature, because many authors use the old β symbol for the feedback ratio.

Carefully examine the circuit shown in Fig. 4-8. This is a simple two-stage amplifier circuit with RC coupling. There is nothing new here. The voltage gain is the ratio of the output voltage (E_o) to the input voltage (E_i):

$$A_v = \frac{V_o}{V_i}$$

Everything is perfectly straightforward.

In Fig. 4-9 we have the same circuit, but with two additional components (C_f and R_f) providing a feedback path from the collector (output) of Q2 back to the emitter of Q1. This immediately complicates the issue of gain, because the feedback signal must be considered. The equation for positive feedback becomes:

$$A_f = \frac{A_v}{1 - BA_v}$$

where A_f is the circuit gain including feedback, A_v is the gain without feedback, and B is the feedback to output ratio (V_f/V_o).

It is generally easier to work with the emitter and feedback

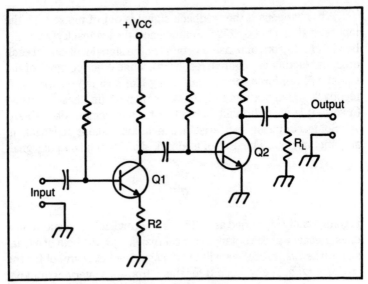

Fig. 4-8. A two-stage amplifier with RC coupling.

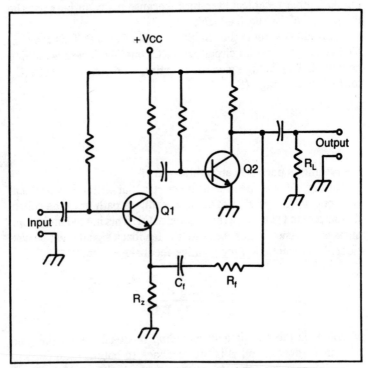

Fig. 4-9. A feedback path added to the circuit of Fig. 4-8.

154

resistances to calculate the feedback ratio (B). The equation becomes:

$$B = \frac{R_e}{R_e + R_f}$$

Let's see how positive feedback can affect the gain of a typical amplifier circuit. Let's assume we have an amplifier with a gain of 100. The emitter resistor has a value of 62 ohms, and the feedback resistor is 4700 ohms. This makes the feedback ratio equal to:

$$B = \frac{62}{62 + 4700} = 0.013$$

In this circuit, when we add the effects of positive feedback, we get an overall circuit gain of:

$$A_f = \frac{A_v}{1 - B_v} = \frac{100}{1 - (0.013 \times 100)}$$

$$= \frac{100}{1 - 1.3} = \frac{100}{-0.3} = -333.33$$

(Ignore the minus sign).

Notice how much the gain has been increased. It wouldn't take much positive feedback to totally overload the transistor.

For negative feedback, the formula is quite similar, but the sign of the operation in the denominator is changed from minus to plus:

$$A_f = \frac{A_v}{1 + BA_v}$$

Using the values from the previous example, the gain with negative feedback works out to:

$$A_f = \frac{100}{1 + (0.013 \times 100)} = \frac{100}{1 + 1.3} = \frac{100}{2.3}$$

$$= 43.48$$

Negative feedback decreases the overall gain.

If the BA_v is considerably larger than 1, the equation can be simplified to:

$$A_f \cong \frac{1}{B}$$

For example, let's work with an amplifier with the following values:

$$A_v = 1500$$
$$R_e = 82 \text{ ohms}$$
$$R_f = 2200 \text{ ohms}$$

The feedback ratio in this case is:

$$B = \frac{82}{82 + 2200} = \frac{82}{2282} = 0.036$$

The final gain with negative feedback is:

$$A_f = \frac{1500}{1 + (0.036 \times 1500)} = \frac{1500}{1 + 54} = \frac{1500}{55} = 27.27$$

Since BA_v is considerably greater than 1 (54), we can use the simpler formula to approximate the gain with feedback:

$$A_f \cong \frac{1}{B} \cong \frac{1}{0.036} = 27.78$$

The result is not precise, but it is close enough for most applications.

Adding feedback to an amplifier circuit has additional effects besides changing the gain. For one thing, the input impedance increases by a factor directly proportional to the feedback ratio. Similarly, the output impedance is reduced, because the feedback signal is tapped off from across the load. The output impedance is reduced by an amount indirectly proportional to the feedback factor.

Negative feedback also reduces distortion, and improves the frequency response of the amplifier.

If the distortion in an amplifier without feedback is D, adding negative feedback will result in the distortion being reduced to:

$$D_f = \frac{D}{B}$$

If the amplifier without feedback has a frequency response ranging from F_L (the low frequency limit) to F_H (the high frequency limit), the addition of negative feedback will extend these limits. The low frequency limit becomes:

$$F_{Lf} = \frac{F_L}{1 + BA_v}$$

And the high frequency limit becomes:

$$F_{Hf} = F_H(1 + BA_v)$$

Let's return to the sample amplifier discussed in our previous examples. The gain without feedback is 100 (A_v), and the feedback factor is 0.013 (B). We will go on to assume that the low frequency limit without feedback (F_L) is 200 Hz, and the high frequency limit without feedback (F_H) is 10000 Hz.

The addition of the negative feedback lowers the low frequency limit to:

$$F_{Lf} = \frac{F_L}{1 + BA_v} = \frac{200}{1 + (0.013 \times 100)}$$

$$= \frac{200}{1 + 1.3} = \frac{200}{2.3} = 86.96 \text{ Hz}$$

Similarly, the high frequency limit is increased to:

$$F_{Hf} = F_H(1 + BA_v = 10000 (1 + (0.013 \times 100))$$
$$= 10000 (1 + 1.3) = 10000 (2.3) = 23000 \text{ Hz}$$

As you can see, negative feedback can have a significant effect on the frequency response of an amplifier.

The circuit shown in Fig. 4-9 uses series feedback. In Fig. 4-10 we have a circuit which uses parallel feedback. The basic concepts and equations are the same.

AMPLIFIER CLASSES

There are several classes of amplifiers. The distinction between

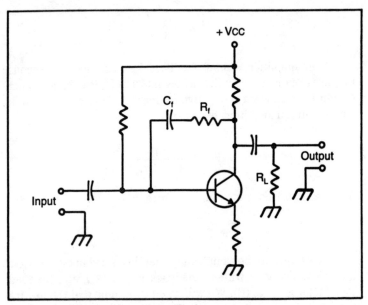

Fig. 4-10. Amplifier using parallel feedback.

the classes depends on the biasing of the transistors used in the circuits.

Ordinarily, the transistor is biased so the quiescent operating point is exactly at the midpoint of the circuit's output range. For example, if the output can swing from 0 to 10 volts, the quiescent operating point will be set at 5 volts. The output can move an equal distance in either direction. The transistor operates linearly throughout its range, as shown in Fig. 4-11.

This is called a *Class A* amplifier. Class A amplifiers are primarily used in low power applications.

It is obvious that to deliver power, a transistor must dissipate power. If the maximum power dissipation rating of the transistor is exceeded at any point on the load line, the transistor will be damaged, or destroyed. The maximum power dissipation rating (usually included in the manufacturer's specification sheet for the transistor) limits the maximum product of the collector to emitter voltage and the collector current.

As an example, let's consider a transistor with a maximum power dissipation rating of 12 watts, and a maximum collector current of 1.5 amperes. Table 4-3 lists the maximum allowable collector to emitter voltages for various values of collector current. These values must never be exceeded, even momentarily. This data is

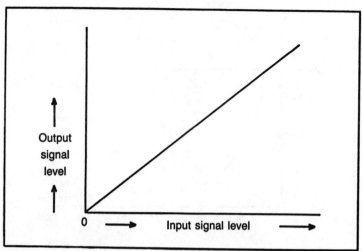

Fig. 4-11. Class A amplification.

often given in the form of a maximum power dissipation graph, like the one shown in Fig. 4-12.

The maximum power dissipation rating puts a fairly low limit on the maximum output power a Class A amplifier can put out.

Table 4-3. A Summary of the Maximum
Collector/Emitter Voltages for a Typical Transistor.

Collector Current	Collector-to-Emitter Voltage
(I_c)	(V_{ce})
0	∞
0.1	120
0.2	60
0.3	40
0.4	30
0.5	24
0.6	20
0.7	17.14
0.8	15
0.9	13.33
1.0	12
1.1	10.91
1.2	10
1.3	9.23
1.4	8.57
1.5 (maximum value)	8

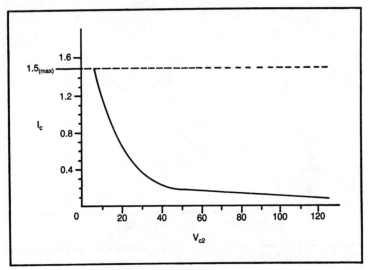

Fig. 4-12. A maximum power dissipation graph.

In addition, a Class A amplifier does not make very efficient use of its power supply. Power is consumed by the circuit, even with no input signal. This can be problematic in portable equipment which is to be powered from small batteries.

Class A amplifiers are often used to power the speakers in portable radios. However, the impedance of most speakers is too low from the transistor to drive directly. An impedance matching transformer is usually used. The turns ratio can be determined by this ratio:

$$\sqrt{(Z_o/Z_s)} : 1$$

where Z_o is the output impedance of the amplifier, and Z_s is the impedance of the load to be driven (the speaker, or whatever). For example, if the output impedance of the amplifier is 95 ohms, and we need to drive an 8 ohm speaker, an impedance matching transformer with the following turns ratio is needed:

$$\sqrt{(95/8)} : 1$$
$$\sqrt{11.875} : 1$$
$$3.445:1 \cong 3.5:1$$

That is, the primary winding of the transformer should have three and a half times more turns than the secondary winding. For ex-

160

ample, if the secondary has 120 turns, the primary should have:

$$120 \times 3.5 = 420 \text{ turns}$$

When an impedance matching transformer is used between the amplifier and the speaker, the efficiency of the circuit as a whole is 50%. That is, the maximum output power is equal to 50% of the power supplied by the power supply (VCC). If we wanted a 10 watt output, the power supply would need to provide at least 20 watts.

Clearly, a Class A amplifier is not a shining example of operating efficiency, and the situation will probably be even worse than it seems at first glance. That 50% efficiency rating is assuming idealized conditions. In a practical Class A amplifier circuit, the overall efficiency may be 20 to 40% lower than the "ideal" 50% rating. That means the actual level of efficiency you can reasonably expect from a Class A amplifier with an impedance matching transformer is 10 to 20%. To get a 10-watt output, the power supply would have to pump in 50 to 100 watts!

If we removed the impedance matching transformer and placed the load directly in the collector circuit, things would be even worse. In this case, we can only expect about 5 to 10% efficiency.

All that wasted power will have to be dissipated by the transistor. This means we will have to use an expensive heavy duty transistor for even a relatively small output.

Still, Class A amplifiers are inexpensive to design and build, which is why they are regularly used in low-end equipment. In addition, they offer relatively low distortion, because the transistor is always operating within its linear region.

Better efficiency can be achieved by using a *Class B amplifier* circuit. In this type of circuit, the transistor is biased so the quiescent current is zero. When there is no input signal, there is no current (or an extremely low current, which may be reasonably ignored).

When the input signal is greater than zero, the transistor conducts and amplifies linearly. When the input signal is zero or less (negative), the transistor is cut off, and the output is zero. The Class B amplifier only passes half of the input signal, as shown in Fig. 4-13.

A Class B amplifier with reversed polarities that conducts during the negative portion of the input signal, but is cut off during the positive portion can also be designed.

A typical Class B amplifier stage is illustrated in Fig. 4-14. It

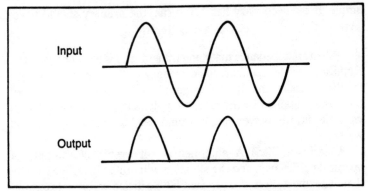

Fig. 4-13. A class B amplifier passes only half the ac input signal.

really isn't all that different from the Class A amplifiers we have already been dealing with. The main difference is in the biasing, and the quiescent operating point.

Cutting off half of the signal would obviously produce an undesirable degree of distortion for most practical applications. The entire signal may be passed by using two complementary Class B stages in parallel, as shown in Fig. 4-15. This is often called a *push-pull* amplifier, because the two transistors essentially divide the labor between themselves. When Q1 "pushes," Q2 "pulls," and vice versa.

The Class B amplifier circuit shown in Fig. 4-15 can put out

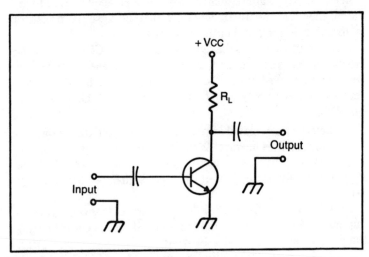

Fig. 4-14. A typical class B amplifier stage.

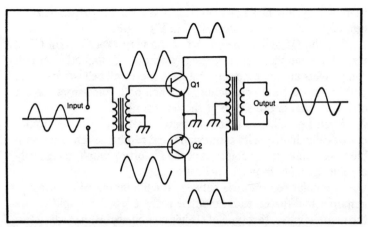

Fig. 4-15. Push-pull amplifier consisting of two class B amplifier stages in parallel.

up to 78.5% of the power supply's input power. That is, if the power supply feeds the circuit 10 watts, the output can be as much as 7.85 watts. Clearly, the Class B amplifier is far more efficient than the Class A amplifier.

Unfortunately, the Class B amplifier has limitations of its own. For one thing, it tends to have a higher amount of distortion than a comparable Class A amplifier.

Obviously, in a single transistor Class B amplifier (like the one in Fig. 4-14) distortion is extremely high, since half the signal is clipped. A push-pull circuit, like Fig. 4-15, has lower distortion, but

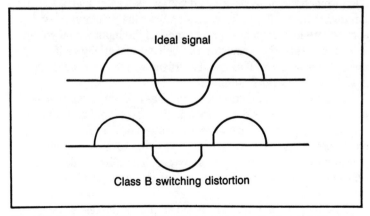

Fig. 4-16. The circuit of Fig. 4-15 tends to produce switching distortion near the zero crossing points in the output signal.

163

there will usually be some distortion of the signal around the zero-crossing point. This is illustrated in Fig. 4-16.

Ideally, Q1 should be cut off at the very exact instant Q2 is turned on, and vice versa. If the two transistors and other circuit components are not precisely matched, there will be instants when both transistors are conducting, or when both transistors are cut off. This affects the shape of the output signal.

It isn't possible to precisely match everything in the circuit, but by using high quality components and careful design, it is possible to minimize the problem, and the distortion won't necessarily be objectionable in most applications.

A popular compromise between the limitations of the Class A amplifier and the Class B amplifier is the *Class AB* amplifier. As the name implies, this type of circuit combines the characteristics of the other two amplifier types.

In a Class A amplifier, the transistor is biased to the center of its output load line. In a Class B amplifier, the transistor is biased so no collector current flows when there is no input signal. Not surprisingly, a Class AB amplifier is biased between those two points.

Some collector current flows when there is no input signal, but this current flow will be less than in a comparable Class A amplifier. This *idling collector current* lowers the efficiency of the amplifier (as compared to a Class B amplifier), but this arrangement also allows lower distortion (as compared to a Class B amplifier).

A Class AB amplifier has less *crossover distortion* than a Class B amplifier. No (or very little) current will flow through a semiconductor junction unless the applied voltage is greater than a specific minimum level. For silicon transistors this minimum level is somewhere in the 0.5 to 0.7 volt range. If the input signal voltage is less than this, the transistor will behave essentially as if the input signal was zero. This can distort the waveform near the zero base line in a Class B amplifier.

Another potential problem in Class B circuits, which is prevented by Class AB operation is that large voltage transients can build up in the circuit when the transistor is rapidly switched on or off. This could cause the transistor to break down, and possibly even be permanently damaged. The idling collector current in a Class AB amplifier minimizes potential problems of this type.

The last of the important basic amplifier classes is the *Class C amplifier*.

A Class C amplifier stage built around an NPN transistor is biased so the base is negative with respect to the emitter when no signal is applied. In other words, the base-emitter junction is normally reverse-biased. The transistor will only conduct for a brief portion of each cycle, when the input signal exceeds the bias voltage. This is illustrated in Fig. 4-17.

Because the transistor only conducts for a small portion of each cycle, the Class C amplifier is capable of putting out very large output signals. This type of circuit offers very high efficiency, up to 85% in many cases.

Figure 4-17 makes it clear that the signal is severely distorted by Class C amplification. Class C amplifiers are almost never used for audio applications. The distortion is unacceptably high. However, Class C amplifiers are often used in rf (radio frequency) amplifiers where the signal frequency and amplitude is important, but the waveshape is noncritical.

Most amplifier circuits are variations of the four basic amplifier classes:

Class A
Class B
Class AB
Class C

In recent years, a number of new amplifier classes have also been developed. We don't have the space to discuss them all in detail here, but we can mention some of the more important ones in passing.

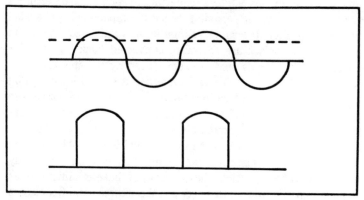

Fig. 4-17. Class C amplification passes only a small portion of the input cycle.

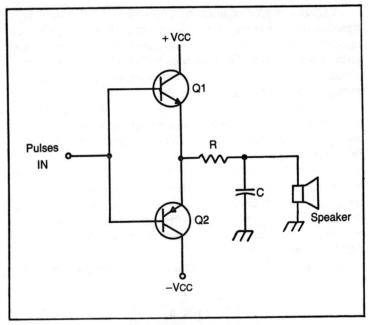

Fig. 4-18. A Class D amplifier amplifies pulsed signals.

Class D amplifiers convert the input signals (usually in the audio range) into pulses. In some circuits, up to 500,000 pulses are produced each second. The width of each pulse is proportional to the instantaneous amplitude of the input signal. The pulses are amplified and passed through a transformerless push-pull amplifier, like the one shown in Fig. 4-18. Resistor R and capacitor C form an integrator (low-pass filter). The width of each pulse determines the voltage across the capacitor. Since the capacitor is in parallel with the speaker, the same voltage is, naturally, fed across the speaker. The pulses are therefore converted back into regular analog (audio) signals, and are reproduced by the speaker.

Class E, Class F, and *Class G amplifiers* are modified push-pull circuits. In each half of the push-pull circuit, a pair of transistors are connected in series. For low input signals, only one of the transistors in each half of the circuit is conducting. The rest of the transistors are turned on when the input signal is high. All the transistors are conducting only when the input signal is high (i.e., only for a portion of each cycle). The efficiency of these amplifier classes is quite high. The differences between these three amplifier classes are fairly subtle, and we don't need to go into them here.

Class H amplifiers are unique in that they use digital logic circuits. Otherwise, they basically resemble standard push-pull circuits. Ordinarily, a Class H amplifier functions like a Class AB amplifier. But when the input signal exceeds a special level, the logic circuitry activates the power supply to feed a higher voltage to the push-pull transistors. Most of the time, the supply voltage and the power dissipation are low. The extra power is only called in when it is needed. Class H amplifiers therefore offer extremely high efficiency.

HIGH FREQUENCY AMPLIFIERS

In our discussions so far, we have been assuming the amplifier circuits are dealing with relatively low (audio) signal frequencies. At relatively low signal frequencies, the problems of circuit design are fairly straightforward. But, once we enter the rf (radio frequency) region, things start to get more complicated. Many factors which can be ignored at low frequencies take on critical importance at higher frequencies.

Probably one of the biggest source of high frequency problems is stray capacitance. Stray capacitance can exist within individual components, or within the wiring of the circuit itself. A capacitor, after all, is simply two conductors separated by an insulator. Any pair of neighboring conductors can behave as a capacitor.

At low frequencies, these stray capacitances are rarely worth mentioning, but, as the signal frequency is raised, they become increasingly important. This is because the reactance (ac resistance factor) of a capacitor is frequency dependent. A capacitor's reactance equals:

$$X_c = \frac{1}{2 \pi FC}$$

where X_c is the capacitive reactance in ohms. π is the mathematical constant pi (approximately 3.14), F is the applied frequency in Hertz, and C is the capacitance in farads.

As an example, let's find the reactance of a 0.1 μF (0.0000001 farad) capacitor at 50 Hz:

$$X_c = \frac{1}{2 \times 3.14 \times 50 \times 0.0000001}$$

$$= \frac{1}{0.0000314} \cong 31{,}831 \text{ ohms}$$

But at 200,000 Hz (200 khz), the capacitive reactance becomes:

$$X_c = \frac{1}{2 \times 3.14 \times 2000000 \times 0.0000001}$$

$$= \frac{1}{0.1256637} \cong 7.96 \text{ ohms}$$

That's certainly a considerable difference. As the applied frequency increases, the capacitive reactance decreases.

Stray capacitances are typically very small. Let's consider a more or less typical example. We'll assume we have two adjacent wires, that should be electrically isolated from each other. But, because of their proximity, a stray capacitance of about 0.000035 μF (0.000000000035 farad) exits between the two wires. Effectively,

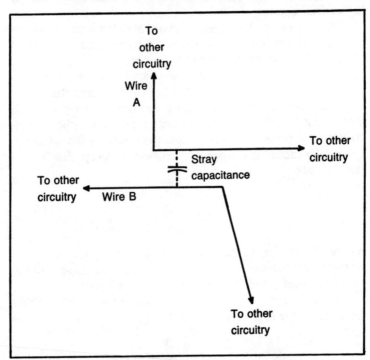

Fig. 4-19. Stray"phantom"capacitances may appear between adjacent wires.

they are connected by a small capacitor, as shown in Fig. 4-19.

At relatively low frequencies, this phantom capacitor will have relatively little effect. For instance, at 300 Hz, the capacitive reactance is:

$$X_c = \frac{1}{2 \times 3.14 \times 300 \times 0.000000000035}$$

$$= \frac{1}{0.000000066} = 15{,}157{,}614 \text{ ohms}$$

Over 15 megohms. This large reactance will allow very little current to flow from wire A to wire B (or vice versa). Effectively, there is almost an open circuit (no connection) between the wires, as there ideally should be.

But let's say that these two adjacent wires are in an rf amplifier, and the signals through them are 4.2 MHz (4,200,000 Hz). In this case, the capacitive reactance drops to:

$$X_c = \frac{1}{2 \times 3.14 \times 4200000 \times 0.000000000035}$$

$$= \frac{1}{0.00009236} \cong 1083 \text{ ohms}$$

With a reactance this low, quite a bit of the signal current can jump from wire to wire. The result is that some of the signals will be in the wrong part of the circuit, affecting circuit operation.

An ideal capacitor would be purely capacitive, but practical components are not ideal. A practical capacitor behaves electrically like an ideal capacitor in parallel with a resistor, as shown in Fig. 4-20. This is a dc resistor (not the ac reactance of the capacitance), and represents the leakage of the capacitor. It will usually have a high value. Some dc can pass through a capacitor, resulting in some degree of signal loss.

Similarly, an ideal inductor (coil) would be purely inductive, with no dc resistance. That is, at dc, the impedance (inductive reactance plus dc resistance) should be zero. Practical inductors behave as though they had a resistor in series with the coil, as shown in Fig. 4-21. Once again, the resistor represents the dc leakage resistance, not the inductive reactance. This phantom resistor will usually have a very small, but non-zero value.

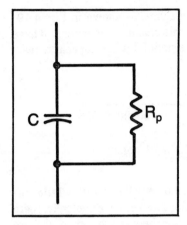

Fig. 4-20. A practical capacitor acts like an ideal capacitance in parallel with a dc resistance.

The better the capacitor, the larger R_p will be. The better the inductor, the smaller R_s will be.

The effect of the leakage resistance in a component or circuit is usually expressed in terms of Q, or *Quality factor*. Q is the ratio between the reactance and the series leakage resistance:

$$Q = \frac{X}{R_s}$$

Or, the ratio between the reactance and the parallel leakage resistance:

$$Q = \frac{R_p}{X}$$

An inductor will also exhibit some capacitive elements, so it will have a parallel leakage resistance, in addition to the series leakage resistance shown in Fig. 4-20. By the same token, a practical capacitor will exhibit some inductance and a series resistance along with the parallel resistance. Either form of the Q equation

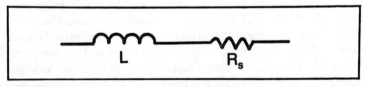

Fig. 4-21. A practical inductor acts like an ideal inductance in series with a dc resistance.

can be used for either type of component, but it is generally more convenient to deal with parallel leakage resistance when dealing with capacitors, and series leakage resistances when dealing with inductors.

We can combine the Q equation with the reactance equation like this:

$$Q = \frac{R_p}{X_c} = \frac{R_p}{\frac{1}{2}\, 2\, \pi\, FC} = 2\, \pi\, FCR_p$$

Let's look at a typical case. Assume the following values:

$$C = 0.001\ \mu F\ (0.000000001\ \text{farad})$$
$$F = 15000\ \text{Hz}$$
$$R_p = 100K = 100,000\ \text{ohms}$$

In this case, the Q of the capacitor would be:

$$Q = 2 \times 3.14 \times 15000 \times 0.000000001 \times 100000 \cong 9.4$$

Earlier we stated that the better the capacitor, the higher the value of R_p will be. If we change the value of R_p to 880K (880,000 ohms), the Q becomes:

$$Q = 2 \times 3.14 \times 15000 \times 0.000000001 \times 880000 \cong 82.9$$

The better the capacitor (the lower the leakage), the higher the Q will be.

If we know the Q, we can solve for the leakage resistance, by rearranging the equation:

$$R_p = \frac{Q}{2\, \pi\, FC}$$

For example, let's say we have a 0.001 μF capacitor with a 15000 Hz signal being placed across it, and the manufacturer's specification sheet says that the Q is 25. The parallel leakage resistance must then be:

$$R_p = \frac{10}{2 \times 3.14 \times 15000 \times 0.0000001}$$

$$= \frac{10}{0.0000942} \cong 106,103 \text{ ohms}$$

The same sort of calculations can be performed with inductors.

Occasionally, the quality factor of a capacitor will be described in terms of DF (Dissipation Factor), instead of Q (quality factor). The *Dissipation Factor* is simply the reciprocal of the quality factor:

$$DF = \frac{1}{Q} = 2 \pi FCR_p$$

For example, when the Q is 25, the DF is:

$$DF = \frac{1}{25} = 0.4$$

The smaller the Dissipation Factor, the better the capacitor.

Stray capacitances and inductances can combine at high frequencies to form undesirable resonances. This can cause instability in an amplifier circuit, and the circuit may even break into oscillation. The concepts of resonance and oscillation will be discussed in detail in the next chapter, so we won't go into them now.

Transistors are also frequency sensitive devices. Many of their operating characteristics can vary with frequency.

For one thing, practical transistors exhibit stray capacitances between their terminals. These stray capacitances are very small, so they are of little significance at low frequencies. But at high frequencies they can have a severe effect on circuit operation. Transistors designed specifically for rf (radio frequency) circuits are manufactured to minimize internal capacitances.

Some inexpensive translators have relatively large internal capacitances, so they can only be used reliably in low frequency circuits.

Throughout this chapter, we have been assuming that alpha (α) and beta (β) have constant (frequency independent) values, but this is actually only true for relatively low signal frequencies. As the signal frequency increases above a specific point, these values become progressively smaller.

Since the input impedance of a transistor circuit is usually related to beta, the input impedance can change with frequency. For example, a typical circuit might have an input impedance equal to:

$$Z_i = \beta \times R_e$$

Since beta gets smaller at high frequencies, so will the input impedance of this circuit.

The drop in alpha and beta with increasing frequency follows a definite pattern. To describe this pattern, we will call the normal, low frequency alpha value α_n. A decrease of 3 dB (decibels) brings the value down to the alpha cutoff frequency (α_c), which relates to α_n by a factor of 0.707:

$$\alpha_c = 0.707 \times \alpha_n$$

The frequency where the alpha equals α_c is given various labels in technical literature. These include:

$$E_\alpha$$
$$F_{hbo}$$
$$F_a$$
$$F_{\alpha b}$$

We will use F_α.

At F_α the alpha is 1/1.4 of its nominal (low-frequency value). At $2F_\alpha$ it drops to 1/2.24 of α_n. At $4F_\alpha$ the alpha is 1/4 of α_n. Above this point, the alpha is halved for each doubling of frequency. For instance, $8F_\alpha$ is twice $4F_\alpha$, so the alpha at this frequency is one-half of what it was for $4F_\alpha$, or 1/8 of α_n. This pattern continues up to extremely high signal frequencies.

This pattern is illustrated in the graph shown in Fig. 4-22. This graph is in logarithmic form. One octave (doubling of frequency) is represented by a constant distance. The alpha values are also shown logarithmically.

The alpha/beta relationship remains constant. This means the beta value will change with the alpha value. The same graph can be used to represent the beta/frequency relationship.

Most transistor specification sheets include a *gain-bandwidth product* value, usually written as F_t. This value can be helpful in finding the beta for any desired frequency. The gain-bandwidth product is defined as the product of beta and the upper 3dB limit of a band of frequencies. For our purposes, the low end of the band is assumed to be 0 Hz (dc).

Beta can be found by dividing the gain-bandwidth product (F_t) by the signal frequency (F_s):

$$\beta = \frac{F_t}{F_s}$$

For example, let's say we have a transistor with a 50 MHz (50,000,000 Hz) gain-bandwidth product (F_t). If we operate the transistor with a 1 MHz (1,000,000 Hz) signal, the beta is:

$$\beta = \frac{50000000}{1000000} = 50$$

If we increase the operating frequency to 2 MHz (2,000,000 Hz), the beta becomes:

$$\beta = \frac{50000000}{1000000} = 25$$

If we decrease the operating frequency to 0.5 MHz (500,000 Hz), the beta becomes:

$$\beta = \frac{50000000}{500000} = 100$$

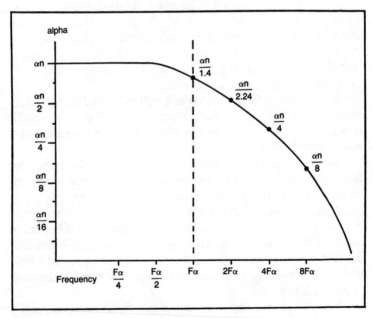

Fig. 4-22. An amplifier does not pass all frequencies equally.

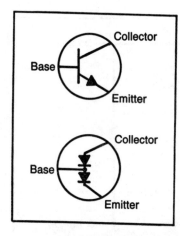

Fig. 4-23. A transistor behaves like two back-to-back diodes at low frequencies.

The functional model (or equivalent circuit) of a transistor also changes at high frequencies. For low frequency operation, a transistor can be represented as two back-to-back diodes, as shown in Fig. 4-23.

This equivalent circuit can be further simplified by replacing the diodes with their appropriate resistances. For ordinary operation, the base-collector diode is reverse-biased, so r_{bc} has a high value. At the same time, the base-emitter diode is forward-biased, so r_{be} has a relatively low value. This simplified equivalent circuit for a transistor is shown in Fig. 4-24.

A transistor also has internal capacitances, of course, but the values are very small, so their effects can be ignored at relatively low frequencies. However, when we start working with high frequencies, we have to add more detail to our equivalent circuit. The

Fig. 4-24. A simplified equivalent circuit for a transistor at low frequencies.

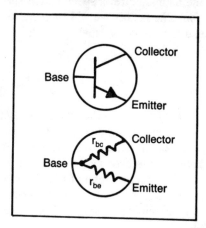

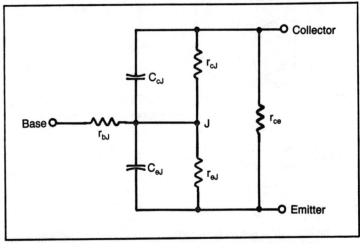

Fig. 4-25. The equivalent circuit for a transistor at high frequencies.

high frequency equivalent circuit of a typical transistor is shown in Fig. 4-25.

Most of the equivalent internal components are referenced to an internal point (J). For example, resistance r_{cJ} is the dc resistance between the collector and point J. Since point J is not directly connected to any of the transistor's leads, we cannot directly measure these values. But these internal values can affect circuit operation, so the circuit designer often needs to know what they are. What's to be done?

Fortunately, these internal equivalent component values can be calculated from other measurements and/or values included in the manufacturer's specification sheet for the transistor in question. The factors we will be using for these calculations generally use the following labels:

$$\beta \qquad h_{oe} \qquad h_{ie}$$
$$g_m \qquad h_{re} \qquad C_{ob}$$

β, of course, is the transistor's beta, which you should already be thoroughly familiar with. It is usually given in the specification sheet, but if not, it can be calculated from the alpha (α), or the ratio between the base and collector currents.

The label g_m refers to the transconductance of the transistor. It specifies the relationship between the collector current (I_c), and the base-emitter voltage (V_{be}). Typically the transconductance is approximately equal to the quiescent (no-input signal) collector current divided by 0.026:

$$g_m = \frac{I_c}{0.026}$$

The collector current is assumed to be in amperes for this equation.

Admittance (often specified by the letter h) is the inverse of resistance. That is, if a component has a resistance of R, it has an admittance of:

$$h = \frac{1}{R}$$

Without a load, the admittance looking back into a transistor (the output admittance) is called h_{oe}. This value will generally be included in the transistor's specification sheet. Since admittance is the reciprocal of resistance, the output resistance of the transistor is:

$$R_o = \frac{1}{h_{oe}}$$

Remember, we are assuming that there is no load on the output. A load impedance will alter these values.

Another common h factor included in transistor specification sheets is h_{ie}. This is the input impedance of the transistor when the output is short-circuited.

The third h factor is h_{re}. This is a ratio between two voltages—the output voltage (V_o), and the voltage at the input, due to voltage present at the output (that is, the feedback voltage) (V_f):

$$h_{re} = \frac{V_f}{V_o}$$

We are again assuming there is no load at the output for this equation.

The factors described so far are for common-emitter circuits. Some manufacturer's specification sheets will list h factors for common-base circuits. We can convert the common-base values to common-emitter values by using the following formulas:

$$h_{oe} = \beta \, h_{ob}$$

$$h_{ie} = Bh_{ib}$$

$$h_{re} = \frac{h_{ie} \times h_{oe}}{\beta} + h_{rb}$$

Finally, we have C_{ob}. This is the collector-to-base capacitance of a transistor's common-base equivalent circuit. This value can be used directly for common-emitter circuit calculations, even though it is based on the common-base circuit. C_{ob} will usually be included in the specification sheet and the transistor.

Once these parameters are known, the values for the internal "components" of the equivalent circuit for a transistor (Fig. 4-25) can be calculated. The equations are as follows:

$$r_{bj} = h_{ie} \qquad r_{eJ} = \frac{\beta}{g_m} \qquad C_{cJ} = C_{ob}$$

$$r_{cJ} = \frac{r_{eJ}}{h_{re}} \qquad 4_{ce} = \frac{1}{h_{oe} - (g_m \times h_{re})} \qquad C_{eJ} = \frac{g_m}{2 \, \pi \, F_t}$$

(F_t, you should recall, is the gain-bandwidth product.)

Table 4-4 shows a portion of a specification sheet for a fairly typical translator. We will use these specified values to calculate the equivalent circuit component values for this transistor.

The transconductance (g_m) is not included on specification sheets, because it's value depends on external circuit values. For our sample problem, we will assume the quiescent collector current (I_c) is 2.5 ma (0.0025 ampere). This makes the transconductance equal to:

Table 4-4. A Portion of a Typical Transistor Specification Sheet.

β	80
F_t	200 MHz (200,000,000 Hz)
C_{ob}	10 pF
h_{oe}	0.00003 (3×10^{-5})
h_{ie}	1250
h_{re}	0.00012 (1.2×10^{-4})

$$g_m = \frac{I_c}{0.026} = \frac{0.0025}{0.026} \quad 0.096 \text{ mhos}$$

(The unit of conductance is the mho, or "ohm" spelled backwards. This is because conductance is the opposite of resistance.)

Now, finding the values for the equivalent circuit is simply a matter of plugging in the appropriate values from the specification sheet:

$$r_{bj} = h_{ie} = 1250 \text{ ohms}$$

$$r_{eJ} = \frac{\beta}{g_m} = \frac{80}{0.096} = 832 \text{ ohms}$$

$$r_{cJ} = \frac{r_{eJ}}{h_{re}} = \frac{832}{0.00012} \cong 6,933,333 \text{ ohms} \cong 6.9 \text{ M}$$

$$r_{ce} = \frac{1}{h_{oe} - (g_m \times h_{re})}$$

$$= \frac{1}{0.00003 - (0.096 \times 0.00012)}$$

$$= \frac{1}{0.00003 - (0.00000115)}$$

$$= \frac{1}{0.0000185} \cong 54,054 \text{ ohms} \cong 54\text{K}$$

$$C_{cJ} = C_{ob} = 10 \text{ pF}$$

$$C_{eJ} = \frac{g_m}{2\pi F_t} = \frac{0.096}{2 \times 3.14 \times 200,000,000}$$

$$= \frac{0.096}{1256637100} \cong 0.000000\,0000\,765$$

$$= 76.5 \text{ pF}$$

These values will have to be considered in the design equations, if the circuit is intended to amplify high frequencies.

EXPERIMENT #5: DESIGNING A PRACTICAL AMPLIFIER

To give you some practical, hands-on experience with amplifier circuit design, we will design and breadboard a practical two stage amplifier circuit.

Most of the parts you will need for this experiment are listed in Table 4-5. The resistance values are not listed here, because they will be calculated in the course of the experiment. Have a variety of common values handy, if possible.

The oscilloscope, frequency counter, and oscillator will be discussed later, when we get to the ac part of this experiment.

Notice that the power supply (VCC) is negative in the diagram. This is because we will be using PNP transistors in our design this time, rather than NPN transistors. As you will see, the design procedures are the same as for an NPN amplifier. Only the polarities are reversed. The positive terminal of the power supply is connected to ground. The negative terminal of the power supply is the VCC connection.

In our discussion of the design of this amplifier, I will assume you will be using a 2N5354 transistor. The specification sheet for this device is given in Table 4-6. If you use a different transistor, check the manufacturer's specification sheet, because some of the calculations may come out differently. This may also be true of

**Table 4-5. A Partial List of the Components
and Equipment needed to Perform Experiment #5.**

2	2N5354 PNP transistor	
1	4.7 ohm resistor	
1	18 ohm resistor	
1	180 ohm resistor	
1	1.8K resistor	(1800 ohms)
1	8.2K resistor	(8200 ohms)
1	10K resistor	(10000 ohms)
1	47K resistor	(47000 ohms)
1	330K resistor	(330,000 ohms)
1	1.8 Meg resistor	(1,800,000 ohms)
3	0.05 μF capacitor	

9 volt voltage source
solderless breadboard
oscilloscope
oscillator
frequency counter (optional)

Table 4-6. the 2N5354 Transistor.

Type	PNP
V_c	25 volts (maximum)
V_{ce}	25 volts (maximum)
V_{eb}	4.0 volts (maximum)
I_c	300 mA (maximum)
POWER DISSIPATION	360 mW (maximum)
F_t	250 MHz
dc beta	100
C_{ob}	8 pF
h_{ie}	1300
h_{oe}	0.000024
h_{re}	0.0001 5
h_{fe}	70 (when I_c = 50 mA)

general purpose replacements, even when the replacement is listed as a direct substitute for the 2N5354.

The circuit we will be using for our design is shown in Fig. 4-26. Take careful note of the polarities! In this circuit we have two cascaded common-emitter stages.

Our ultimate goal in this experiment will be to design and breadboard an audio frequency amplifier with a voltage gain of 300. If each stage has a gain of 20, the total nominal gain will be 400:

$$A_{vt} = A_{v1} \times A_{v2} = 20 \times 20 = 400$$

The circuit gain will be reduced by the negative feedback through C4 and R7, so it is a good idea to overestimate the gain initially.

We will assume the input voltage will range from -10 mV to +10mV. The output voltage will therefore need to swing from -4 volts to +4 volts. A 9-volt power source would allow output swings from -4.5 volts to +4.5 volts in Class A operation, so we will assign a value of -9 volts to Vcc. (This voltage is negative, because the positive terminal is grounded for correct biasing of the PNP transistors.)

Now, where do we begin to design the amplifier? The answer is to break the circuit down into simpler subcircuits. We will start with the second stage, which is shown separately in Fig. 4-27. At this point we will not use the capacitors, since they aren't needed yet, and it is easier to test our work with dc voltages. Also, the feedback path will be omitted for now.

R8 is the load resistor (R_L), R4 is the base resistor (R_b), R5 is

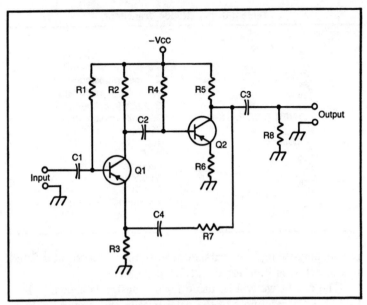

Fig. 4-26. The circuit used for Experiment #5.

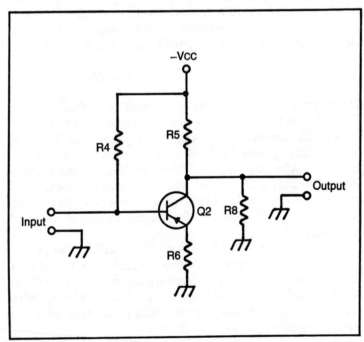

Fig. 4-27. The first part of Experiment #5.

182

the collector resistor (R_c), and R6 is the emitter resistor (R_e).

We will assume that our output load resistance should be about 50,000 ohms (50K). A 47K (47,000 ohms) resistor is a close standard value that can be used for R8.

The collector resistor (R5) should have a value between 10% and 25% of the load resistance. In this case, the range of acceptable values runs from 4.7K (4700 ohms) to approximately 12K (12000 ohms). We will use a 10K (10000 ohm) resistor for R5.

The total external resistance in the collector circuit is the parallel combination of R5 and R8:

$$R_{cc} = \frac{R5 \times R8}{R5 + R8} = \frac{10000 \times 47000}{1000 + 47000} = \frac{470000000}{57000}$$

$$\cong 8246 \text{ ohms}$$

The maximum resistance in the emitter circuit (R_{ee}) for a gain of 25 works out to:

$$R_{ee} = \frac{R_{cc}}{A_v} = \frac{8246}{25} \cong 330 \text{ ohms}$$

Next, we find the collector current (I_c), using Ohm's Law:

$$I_c = \frac{V_c}{R_c} = \frac{4.5}{10000} = 0.00045 \text{ amps} = 0.45 \text{ mA}$$

(The collector voltage (V_c) is biased so that it is the midpoint of the supply voltage, to allow output swings of equal strength for either polarity.)

We can now use the value of I_c to find the internal emitter resistance (r_e):

$$r_e = \frac{26}{I_c} = \frac{26}{0.45} \cong 58 \text{ ohms}$$

Subtracting this from the total allowable emitter resistance (R_{ee}), we come up with the maximum value for the emitter resistor (R6):

$$R6 = R_e = R_{ee} - r_e = 330 - 58 = 272 \text{ ohms}$$

We can leave ourselves plenty of leeway by using a 180 ohm resistor for R6.

To get the desired collector current (0.45 mA), we must divide this value by the dc beta to find the appropriate value for the base current (I_b):

$$I_b = \frac{I_c}{\beta} = \frac{0.45}{100} = 0.0045 \text{ mA} = 0.0000045 \text{ amperes}$$

The next step is to find the total base resistance (R_{bt}), using Ohm's Law:

$$R_{bt} = \frac{V_{CC}}{I_b} = \frac{9}{0.000000045} = 2000000 \text{ ohms}$$

The total base circuit resistance is made up of r_e, R_e (R6), and R_b (R4). We already know two of these resistance values, so it is a simple matter to find the third:

$$R_b = R_{bt} - r_e - R_e = 2000000 - 58 - 180$$
$$= 199762 \text{ ohms}$$

The nearest standard resistance value is 1.8 M (1800000 ohms), so we will use that value for R4.

Now that we have determined values for each of the components, breadboard the circuit of Fig. 4-27. The component values we came up with are shown in Table 4-7 for your convenience.

Also breadboard the dc input circuit shown in Fig. 4-28. This is a simple variable voltage divider network. Use the following resistor values:

Table 4-7. Calculated Parts Values for the Subcircuit of Fig. 4-27.

Q2	2N5354
Vcc	9 volts (watch the polarity!)
R4	1.8 Meg (1,800,000 ohms)
R5	10K (10,000 ohms)
R6	180 ohms
R8	47K (47,000 ohms)

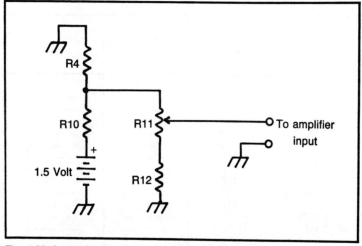

Fig. 4-28. Input circuit to input dc voltages for Experiment #5.

$$R9 = 1K\,(1{,}000\ \text{ohms})$$
$$R10 = 10K\,(10{,}000\ \text{ohms})$$
$$R11 = 10K\ \text{potentiometer}\,(10{,}000\ \text{ohms})$$
$$R12 = 470\ \text{ohms}$$

Using these values, the input voltage to the amplifier can range from 0.006 volt to 0.136 volt. Assuming we have achieved our desired gain of 25, the output voltage with these inputs should range from 0.15 volt to 3.4 volts.

Using a sensitive dc voltmeter, set the input circuit for 0.01 volt (10 mV), then move the voltmeter to the output of the amplifier. Write your result down in Table 4-8. Repeat this procedure for each of the values listed in the table.

Now compare your results to the ideal values listed in Table 4-9. How close did you come? There will almost definitely be some variation between your results and the ideal values listed in Table 4-9. This is because of the component tolerances, and the rounding off of values in the calculations. Remember, we substituted the nearest available resistance values for the calculated values. Naturally, this will have an effect on the output.

Divide each of your output measurements by your input measurements to find the actual gain of your circuit. An extra column is provided in Table 4-8 for this purpose. While there may be some small variation, the gain should remain constant for each input/output combination. If there are significant differences, you

Table 4-8. Worksheet for Part 1 of Experiment #5.

Input Voltage	(mV)	Output Voltage	Gain
0.010	(10)	————	————
0.020	(20)	————	————
0.030	(30)	————	————
0.040	(40)	————	————
0.050	(50)	————	————
0.060	(60)	————	————
0.070	(70)	————	————
0.080	(80)	————	————
0.090	(90)	————	————
0.100	(100)	————	————
0.110	(110)	————	————
0.120	(120)	————	————

probably made an error in your measurements. Repeat the questionable measurements.

We can also doublecheck our results mathematically with the standard resistor values we selected. The actual stage gain is dependent on the external resistance in the collector circuit (R_{cc}), and the resistance in the emitter circuit (R_{ee}):

Table 4-9. Compare Your Results in Table 4-8 to
These Ideal Output Values for the Input Voltages.

Input Voltage	(mV)	Output Voltage	(mV)
0.010	(10)	0.25	(250)
0.020	(20)	0.50	(500)
0.030	(30)	0.75	(750)
0.040	(40)	1.00	(1000)
0.050	(50)	1.25	(1250)
0.060	(60)	1.50	(1500)
0.070	(70)	1.75	(1750)
0.080	(80)	2.00	(2000)
0.090	(90)	2.25	(2250)
0.100	(100)	2.50	(2500)
0.110	(110)	2.75	(2750)
0.120	(120)	3.00	(3000)

$$A_v = \frac{R_{cc}}{R_{ee}}$$

R_{cc} is the parallel combination of the collector resistor (R5) and the load resistor (R8). We have already calculated this for our selected resistance values. The total collector circuit resistance works out to approximately 8246 ohms.

R_{ee} is the sum of the internal emitter resistance (r_e), and the external emitter resistor (R6):

$$R_{ee} = r_e + R6 = 58 + 180 = 238 \text{ ohms}$$

So the voltage gain takes a nominal value of:

$$A_v = \frac{8246}{238} = 35$$

That is higher than our nominal goal of a stage gain of 20. But the gain is brought down due to the effects of the input impedance (or, input resistance for dc).

Because of our resistive input network (Fig. 4-28), we cannot precisely define the input source impedance (resistance). It varies with the input level (different settings of potentiometer R11). But we can estimate, and say that is about 4.5K (4,500 ohms).

The input impedance of the transistor circuits works out to:

$$R_i = \frac{R4 \times (R6 + r_e)}{R4 + (R6 + r_e)} = \frac{1800000 \times 100 \times (180 + 58)}{1800000 + 100 \times (180 + 58)}$$

$$= \frac{1800000 \times 100 \times (238)}{1800000 + 100 \times (238)} = \frac{1800000 \times 23800}{1800000 + 23800}$$

$$= \frac{42840000000}{1823800} \cong 23489 \text{ ohms}$$

Combining this value with the source impedance (R_g), we find the input gain drop factor is equal to:

$$\frac{R_i}{R_i + R_g} = \frac{23489}{23489 + 4500} \cong 0.84$$

which means that approximately 84% of the input signal makes it through to the transistor for amplification. The adjusted voltage gain is therefore:

$$A_{va} = A_v \times 0.84 = 35 \times 0.84 = 29.4$$

That still gives us some extra gain to allow for any additional losses in the circuit. It is better for the gain to be too high than too low. The final output gain can always be reduced by a simple volume control (potentiometer) if necessary.

Your experimental gains listed in Table 4-8 should be somewhat higher than the nominal design value of 20.

In our final amplifier, however, this stage will be driven not by the dc voltage divider circuit of Fig. 4-28. Instead, we will be driving this stage with the output signal from another common-emitter amplifier stage. The output impedance of a common-emitter amplifier stage is considerably higher than the value we used for R_g. A practical estimate would be in the neighborhood of 50K (50,000 ohms). Let's redo the last two equations to see how this effects the gain of this stage.

First, the input gain drop factor is now:

$$\frac{23489}{23489 + 5000} = \frac{23489}{73489} = 0.32$$

Therefore, the adjusted voltage gain of this stage drops to:

$$A_{va} = 35 \times 0.32 = 11.2$$

That is considerably lower than our desired value of 20.

We could compensate by increasing the gain of the first stage, but it would be a better idea to first go back and boost the gain of this stage by reducing the value of the emitter resistor (R6). Circuit design often involves trying out one set of values, then going back and adjusting those values to attempt to come closer to the desired results. This is one reason why a good breadboarding system is so important to the experimenter. Sometimes things will look good on paper, but won't behave as expected when physically put to work with practical components.

Reducing the value of the emitter resistor (R_e or R6) will reduce the value of the input impedance (R_i), so the input gain factor will be increased. Therefore, to make a significant change in the gain, we must change the emitter resistance quite a bit.

We will try dropping the value of R6 by a factor of ten. Instead of 180 ohms, it will now be 18 ohms. Let's see how this will work out. This stage's input impedance (R_i) takes a new value of:

$$R_i = \frac{1800000 \times 100 \times (18 + 58)}{1800000 + 100 \times (18 + 58)}$$

$$= \frac{1800000 \times 100 \times (76)}{1800000 + 100 \times (76)} = \frac{1800000 \times 7600}{1800000 + 7600}$$

$$= \frac{13680000000}{1807600} = 7568 \text{ ohms}$$

At the same time, the nominal gain is increased to:

$$A_v = \frac{8246}{18 + 58} = \frac{8246}{76} = 108.5$$

This is so much higher than our desired value (20) that we've got to have enough gain now, even accounting for the input gain drop factor, which is now:

$$\frac{7568}{7568 + 50000} = \frac{7568}{57568} = 0.13$$

So the effective stage gain has become equal to:

$$A_{va} = 108.5 \times 0.13 = 14.1$$

Good grief! It's still too low! And we can't make R6 much smaller. The only thing we can do is to compensate by increasing the gain of the first stage, or adding another stage to our complete circuit.

Why did we run into this problem? We are working with a common-emitter to common-emitter cascade. The output impedance of a common-emitter amplifier stage is moderately high, but the input impedance of a common-emitter amplifier stage is moderately low. We have something of an impedance mismatch here. This results in inefficient power transfer between stages. Earlier in this chapter we stated that the common-emitter to common-emitter type of cascade gives only acceptable results. In some cases, such "acceptable" results may not be sufficient. But we can live with it here.

We could add a third common-collector stage between the two common-emitter stages, but that will increase the complexity of the circuit as a whole. We can simply increase the gain of the first stage to compensate for the losses of the second stage due to the impedance mismatch.

How much gain should we give the first stage? Well, our goal is to design an amplifier with an overall gain of 300. To allow for various "fudge" factors and the negative feedback (to be added later), we boosted this to 400. The gain of a multi-stage amplifier equals:

$$A_{vt} = A_{v1} \times A_{v2}$$

This can be rearranged to find the necessary gain for stage 1:

$$A_{v1} = \frac{A_{vt}}{A_{v2}}$$

Our calculations gave us a nominal gain for stage 2 of 14.1. Let's reduce this further to allow for more fudge factors. We'll assume component tolerances will drop the gain from its nominal value. Remember, it's usually better to have too much gain, rather than too little. It's next to impossible to work out a design that will hit the desired value right on the nose.

Therefore, we will say the gain of the second stage is only 12. This means the gain of the first stage must be at least:

$$A_{v1} = \frac{400}{12} = 33.33$$

Let's boost that to a nice even 35 for our nominal gain for stage 1. That is not an unreasonable amount of gain.

The components involved with the first stage of our amplifier circuit are shown in Fig. 4-29. We will first breadboard this section of the circuit independently, then combine it with the second stage to create our complete amplifier. We are still ignoring the feedback components (C4 and R7). Coupling capacitors are also omitted for the time being.

Remember to watch the supply voltage polarity. We are work-

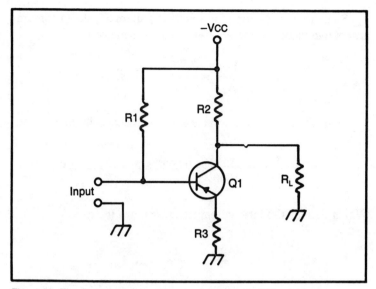

Fig. 4-29. The first stage from the complete circuit of Fig. 4-26.

ing with a PNP transistor, so VCC is negative with respect to ground.

Resistor R_L is a temporary load resistor, used to represent the input impedance of the second amplifier stage, which we have already designed. The input impedance (R_i) of the second stage is approximately 7568 ohms. We will use a 8.2K (8200 ohm) resistor, because that is the closest readily available standard resistor value.

VCC, of course, is still 9 volts, since the same power supply is used to drive both stages in the circuit.

Q1 is a 2N5354 transistor, identical to Q2, so we will still be using the same specification sheet, which is shown in Table 4-6.

Let's quickly run through the design calculations for this stage. You should be fairly well familiar with the procedure by now.

Since the load resistor is 8.2K, a good value for the collector resistor (R2) would be 1.8K (1800 ohms) (22% of R_L).

The total external resistance in the collector circuit (R_{cc}) works out to:

$$R_{cc} = \frac{R2 \times R_L}{R2 + R_L} = \frac{1800 \times 8200}{1800 + 8200} = \frac{14760000}{10000}$$

$$= 1476 \text{ ohms}$$

Since we are shooting for a voltage gain of 35, the maximum acceptable value for the emitter circuit resistance is:

$$R_{ee} = \frac{R_{cc}}{A_v} = \frac{1476}{35} = 42 \text{ ohms}$$

Ohm's Law gives us the value of the quiescent collector current:

$$I_c = \frac{V_c}{R_c} = \frac{4.5}{1800} = 0.0025 \text{ amps} = 2.5 \text{ mA}$$

Which defines the internal emitter resistance (r_e) as:

$$r_e = \frac{26}{I_c} = \frac{26}{2.5} = 10.4$$

So the maximum value for the external emitter resistor (R3) is:

$$R3 = R_{ee} - r_e = 42 - 10.4 = 31.6 \text{ ohms}$$

Let's leave ourselves plenty of elbow room, and assign a value of 18 ohms to R3. This makes the nominal stage gain equal to:

$$A_v = \frac{R_{cc}}{R_{ee}} = \frac{R_{cc}}{R3 + r_e} = \frac{1476}{18 + 10.4}$$

$$= \frac{1476}{28.4} \cong 52$$

The base current is calculated from the transistor's beta, and the collector current:

$$I_b = \frac{I_c}{\beta} = \frac{0.0025}{100} = 0.000025 \text{ ampere}$$

$$= 0.0025 \text{ mA} = 25 \text{ } \mu A$$

This value calls for a total base circuit resistance (R_{bt}) of:

$$R_{bt} = \frac{Vcc}{I_b} = \frac{9}{0.0000025} = 360000 \text{ ohms} = 360K$$

Removing the emitter circuit resistances (R3 and r_e) doesn't have a very significant effect on this value:

$$R1 = R_{bt} - R3 - r_e = 360000 - 18 - 10.4$$
$$= 359971.6 \text{ ohms}$$

The closest standard resistor value is 330K (330000 ohms), so that is what we'll use.

Breadboard this circuit using the component values listed in Table 4-10. Re-use the dc input circuit from the first part of this experiment. This subcircuit was shown in Fig. 4-24. Take output measurements for various input levels, as you did with the first part of this experiment. Record your results in Table 4-11, then compare them to the calculated values listed in Table 4-12.

The input impedance for this stage works out to:

$$R_i = \frac{R1 \times (R3 + r_e)}{R1 + (R3 + r_e)} = \frac{330000 \times 100 \times (28.4)}{330000 \times 100 \times (28.4)}$$

$$= \frac{330000 \times 2840}{330000 + 2840} = \frac{937200000}{332840} = 2816 \text{ ohms}$$

If we assume a source impedance (R_g) of 4500 ohms, we get an input gain drop factor of:

$$\frac{2816}{2816 + 4500} = \frac{2816}{7316} = 0.385$$

Table 4-10. Calculated Parts Values for the Subcircuit of Fig. 4-29.

Q1	2N5354
Vcc	9 volts (watch the polarity)
R1	330K (330,000 ohms)
R2	1.8K (1800 ohms)
R3	18 ohms
R_L	8.2K (8200 ohms)

Table 4-11. Worksheet for Part 2 of Experiment #5.

Input Voltage	(mV)	Output Voltage	Gain
0.010	(10)	———	———
0.020	(20)	———	———
0.030	(30)	———	———
0.040	(40)	———	———
0.050	(50)	———	———
0.060	(60)	———	———
0.070	(70)	———	———
0.080	(80)	———	———
0.090	(90)	———	———
0.100	(100)	———	———
0.110	(110)	———	———
0.120	(120)	———	———

So the stage gain is dropped down to:

$$A_{va} = 52 \times 0.385 = 20$$

If we combine this with the second stage gain, we get a total circuit gain of:

$$A_{vt} = A_{v1} \times A_{v2} = 20 \times 14.1 = 282$$

That's a little low. We could possibly squeak by with this, but let's try to increase the gain a little before we give up. If we replace the emitter resistance (R3) with a 4.7 ohm unit, the nominal gain becomes:

Table 4-12. Compare Your Results
in Table 4-11 to These Calculated Values.

Input Voltage	(mV)	Output Voltage
0.010	(10)	0.200
0.020	(20)	0.400
0.030	(30)	0.600
0.040	(40)	0.800
0.050	(50)	1.000
0.060	(60)	1.200
0.070	(70)	1.400
0.080	(80)	1.600
0.090	(90)	1.800
0.100	(100)	2.000
0.110	(110)	2.200
0.120	(120)	2.400

$$A_v = \frac{R_{cc}}{R3 + r_e} = \frac{1476}{4.7 + 10.4} = \frac{1476}{15.1}$$

$$= 97.75$$

This makes the input impedance equal to:

$$R_i = \frac{330000 \times 100 \times (15.1)}{330000 + 100 \times (15.1)} = \frac{330000 \times 1510}{330000 + 1510}$$

$$= \frac{498300000}{331510} = 1503 \text{ ohms}$$

The input gain drop factor is now:

$$\frac{1503}{1503 + 4500} = \frac{1503}{6003} = 0.25$$

This drops the stage gain to:

$$A_{va} = 97.75 \times 0.25 = 24.44$$

Combining this with the second stage gain we get a total circuit gain (ignoring feedback) of:

$$A_{vt} = A_{v1} \times A_{v2} = 14.1 \times 24.44 = 344.604$$

That's about 15% over our ultimate target gain of 300 (including feedback), although it is less than the 400 without feedback gain we were shooting for. Still, it is close enough.

A lot depends on the source impedance of the input signal. For example if its source impedance is 600 ohms, the equations work out like this:

$$\text{input gain drop factor} = \frac{1503}{1503 + 600} = \frac{1503}{2103}$$

$$= 0.715$$

$$A_{va} = 97.75 \times 0.715 = 70$$

$$A_{vt} = 70 \times 14.1 = 987$$

Considerably more than we need:

But if the signal source has a high source impedance, say 10K (10000 ohms), we get:

$$\text{input gain drop factor} = \frac{1503}{1503 + 10000}$$

$$= \frac{1503}{11503} = 0.131$$

$$A_{va} = 97.75 \times 0.0131 = 12.8$$

$$A_{vt} = 12.8 \times 14.1 = 180.48$$

Much less than we were shooting for.

In one sense, circuit design is not really very difficult. It is simply a matter of working your way through a series of steps, which individually are fairly simple. On the other hand, circuit design can be very complex, because the real world just won't cooperate with your equations. There are many interacting factors to be considered, but often you don't know what they are until after you've done a great deal of work on the preliminary design.

In hindsight, we probably should have used a transistor with a higher dc current gain (beta). But, for the purposes of our experiment, we will stick to the 2N5354. We will assume the original value of 4500 ohms for the input source impedance (R_g). This gives us a combined gain of about 344 for the two stages. We will lower our expectations somewhat, and accept a total circuit gain of 250 when we add the negative feedback.

MORE ON EXPERIMENT #5: AC SIGNALS

When we first began this experiment, we stated that our goal was to build an audio frequency amplifier. So far we have only been working with dc amplifier stages. The input signals we will be interested in are ac signals with frequencies in the 20 kHz to 20 Hz (20000 Hz) range.

Disconnect the dc input voltage divider network (Fig. 4-29) from your breadboarded circuit, but leave the transistor amplifier stage we just designed (Fig. 4-30).

For ac operation we need isolation from the dc bias voltages. Therefore, add an input and output capacitor to the circuit, as il-

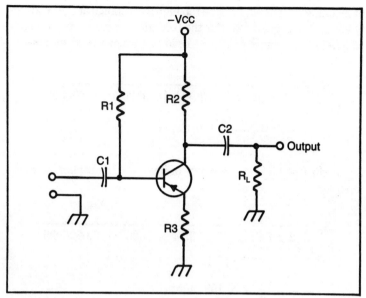

Fig. 4-30. The circuit for part 4 of Experiment #5.

lustrated in Fig. 4-31. Use 0.05 μF capacitors. (They may be marked 0.047 μF, which is the same value, for all intents and purposes.)

How did we select the capacitor values? Actually, it is somewhat arbitrary, but we need a value that will do a reasonably good job

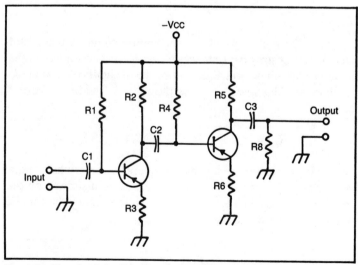

Fig. 4-31. The circuit for Experiment #5 without the feedback components.

of blocking dc, but won't have too high a reactance at the relatively low audio frequencies. For the 0.05 μF capacitors recommended here, the reactance at 20 Hz (the lowest frequency of interest) is:

$$X_c = \frac{1}{2 \pi FC} = \frac{1}{2 \times 3.14 \times 20 \times 0.00000005}$$

$$= \frac{1}{0.0000063} = 159,155 \text{ ohms}$$

At 20 khz (the highest frequency of interest), the capacitive reactance is:

$$X_c = \frac{1}{2 \pi FC} = \frac{1}{2 \times 3.14 \times 20000 \times 0.0000005}$$

$$= \frac{1}{0.006283} = 159 \text{ ohms}$$

Remember, the capacitive reactance will combine with the dc resistances to effect circuit operation at various frequencies. For example, let's consider the nominal gain equation:

$$A_v = \frac{R_{cc}}{R_{ee}}$$

R_{ee}, the total circuit resistance in the emitter circuit, is a constant value (ignoring stray capacitances within the transistor and in the circuit wiring—these will almost never be of significance at audio frequencies). This means R_{ee} has the value it had in our earlier equations.

$$R_{ee} = R3 + r_e = 4.7 + 10.4 = 15.1$$

For R_{cc} (collector circuit resistance), however, we have a different story. The capacitive reactance of C2 is in series with the load resistance R_L. The equation for R_{cc} at dc is simply the parallel combination of R2 and R_L:

$$R_{cc} = \frac{R2 \times R_L}{R2 + R_L}$$

For ac signals, this formula becomes:

$$R_{cc} = \frac{R2 \times (R_L + X_c)}{R2 + (R_L + X_c)}$$

At 20 Hz the capacitive reactance is 159155 ohms, so the total external collector circuit impedance is:

$$R_{cc} = \frac{1800 \times (8200 + 159155)}{1800 + (8200 + 159155)}$$

$$= \frac{1800 \times 167355}{1800 + 167355}$$

$$= \frac{1372311000}{175555}$$

$$= 7817 \text{ ohms}$$

So the nominal gain at 20 Hz is approximately:

$$A_v = \frac{7187}{15.1} = 518$$

The dc gain, you will remember, was 97.75.

Now, if we raise the signal frequency to 20000 Hz, the capacitive reactance is 159 ohms, so R_{cc} becomes:

$$R_{cc} = \frac{1800 \times (8200 + 159)}{1800 + (8200 + 159)} = \frac{1800 \times 8359}{1800 + 8359}$$

$$= \frac{15046200}{10159} = 1481 \text{ ohms}$$

This makes the nominal gain at 20000 Hz about:

$$A_v = \frac{1481}{15.1} = 98$$

Capacitive coupling affects frequency response.

To test our amplifier stage with ac signals, we will need some additional equipment. An oscillator is an ac signal source. Most commercial breadboarding systems include one or more oscillator circuits.

If you do not have a pre-built oscillator circuit available, you can use one of the circuits presented in the next chapter.

The oscillator you use should allow you to adjust the signal frequency, and the output level. Many oscillators have a fixed output signal level. If this is the case, drop the signal voltage with a voltage divider network. The simplest approach would be to use the simple resistive voltage divider circuit shown in Fig. 4-28. Simply replace the battery with the oscillator output. This is not the best approach, because the varying resistances will affect the source impedance seen by the amplifier, but it will do for our quick and dirty tests in this experiment.

An oscilloscope is virtually mandatory for any kind of ac circuit. A simple VOM just won't do the job. You can substitute an ac voltmeter for the oscilloscope if you use only sine wave signals (described in Chapter 5), but the results will not be as informative as with an oscilloscope.

A frequency counter would also be helpful , but it's not absolutely essential. The frequency of the signal can be estimated with the oscilloscope display. A frequency counter simply makes the task a little more convenient.

I will assume you know how to operate these pieces of test equipment. If not, there are many fine books that will introduce you to their functions and operation.

Connect the oscilloscope output to the input of the circuit shown in Fig. 4-30. Use the frequency counter and/or oscilloscope to adjust the peak input voltage to 0.05 volt (50 mV). This same voltage can also be described in the following ways:

PEAK-TO-PEAK—0.1 volt (100 mV)

AVERAGE—0.0318 volt (31.8 mV)

RMS—0.03535 volt (35.35 mV)

These are all different ways for expressing the same voltage. For the rest of this experiment, we will stick with peak voltage values.

The frequency of the oscillator should be set to 100 Hz.

Once the input signal has been set (100 Hz, 0.05 volt peak), move the oscilloscope probe to the output of the amplifier circuit.

Table 4-13. Worksheet for Part 3 of Experiment #5.

	(All voltages are Peak ac Voltages)			
Signal Frequency	**Input Voltage**	**(mV)**	**Output Voltage**	**Gain $\dfrac{V_c}{V_I}$**
100	0.05	(50)	————	————
100	0.10	(100)	————	————
100	0.15	(100)	————	————
500	0.05	(50)	————	————
500	0.15	(100)	————	————
500	0.15	(150)	————	————
1000	0.05	(50)	————	————
1000	0.10	(100)	————	————
1000	0.15	(150)	————	————
2000	0.05	(50)	————	————
2000	0.10	(100)	————	————
2000	0.15	(150)	————	————
4000	0.05	(50)	————	————
4000	0.10	(100)	————	————
4000	0.15	(150)	————	————
6000	0.05	(50)	————	————
6000	0.10	(100)	————	————
6000	0.15	(150)	————	————
8000	0.05	(50)	————	————
8000	0.10	(100)	————	————
8000	0.15	(150)	————	————
10000	0.05	(50)	————	————
10000	0.10	(100)	————	————
10000	0.15	(150)	————	————

The output should have the same waveshape and frequency as the original input signal, but with a higher peak voltage level. Write the peak voltage of the output signal in the space provided in Table 4-13.

Repeat the procedure for each of the input signals listed in Table 4-13. Compare your results to the calculated values given in Table 4-14. Expect some variation between these nominal values and your results due to component tolerances and small errors in measurement. Your results should show the same basic pattern indicated by Table 4-14 however.

Now, let's check the combined gain of the two amplifier stages we have designed. Breadboard the circuit shown in Fig. 4-31, using the component values listed in Table 4-15. Use the oscillator as your input signal source, as in the last portion of this experiment. A worksheet for this part of the experiment is listed in Table

Table 4-14. Calculated Values for Part 3 of Experiment #5.

(All Voltages are Peak ac Voltages)			
Signal Frequency	Input Voltage	Output Voltage	Gain
100	0.05	1.425	28.5
100	0.10	2.850	28.5
100	0.15	4.275	28.5
500	0.05	1.325	26.5
500	0.10	2.650	26.5
500	0.15	3.975	26.5
1000	0.05	1.290	25.8
1000	0.10	2.580	25.8
1000	0.15	3.870	25.8
2000	0.05	1.265	25.3
2000	0.10	2.530	25.3
2000	0.15	3.795	25.3
4000	0.05	1.245	24.9
4000	0.10	2.490	24.9
4000	0.15	3.735	24.9
6000	0.05	1.235	24.7
6000	0.10	2.470	24.7
6000	0.15	3.705	24.7
8000	0.05	1.235	24.7
8000	0.10	2.470	24.7
8000	0.15	3.705	24.7
10000	0.05	1.230	24.6
10000	0.10	2.460	24.6
10000	0.15	3.690	24.6

4-16. Notice that the input voltages specified here are considerably less than in the preceding section. This is because of the much higher gain obtained with the two stages cascaded.

Compare your results with the calculated values listed in Table 4-17.

Be careful when performing this part of the experiment not to feed too high a voltage to the input of the amplifier. The transistors

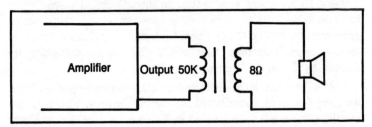

Fig. 4-32. Impedance matching circuit used to allow the circuit of Fig. 4-26 to drive a loudspeaker.

Table 4-15. Component Values Used in Part 4 of Experiment #5.

Q1, Q2	2N5354
R1	330K (330,000 ohms)
R2	1.8K (1800 ohms)
R3	4.7 ohms
R4	1.8 Meg (1,800,000 ohms)
R5	10K (10,000 ohms)
R6	18 ohms
R8	47K (47,000 ohms)
C1, C2, C3	0.05 μF capacitor

could be damaged by the excessive power dissipation through them.

Notice how the gain varies with frequency. For the experimental values we used, the overall gain ranges from 351 for a 10 kHz (10000 Hz) input signal to 438 for a 100 Hz input signal. High audio frequencies won't be amplified as much as lower frequencies. Ideally, all frequencies in the desired range of inputs should be amplified equally.

In addition, all the gains here are greater than our targeted value, so we can add some negative feedback. This is done by ad-

Table 4-16. Worksheet for Part 4 of Experiment #5.

	(All Voltages are in ac Peak)				
Signal Frequency	Input Voltage	(mV)	Output Voltage	Gain	$\dfrac{A_{vo}}{A_{vi}}$
100	0.0050	(5)	————	————
100	0.0075	(7.5)	————	————
100	0.0100	(10)	————	————
500	0.0050	(5)	————	————
500	0.0075	(7.5)	————	————
500	0.0100	(10)	————	————
1000	0.0050	(5)	————	————
1000	0.0075	(7.5)	————	————
1000	0.0100	(10)	————	————
2000	0.0050	(5)	————	————
2000	0.0075	(7.5)	————	————
2000	0.0100	(10)	————	————
4000	0.0050	(5)	————	————
4000	0.0075	(7.5)	————	————
4000	0.0100	(10)	————	————
8000	0.0050	(5)	————	————
8000	0.0075	(7.5)	————	————
8000	0.0100	(10)	————	————
10000	0.0050	(5)	————	————
10000	0.0075	(7.5)	————	————
10000	0.0100	(10)	————	————

**Table 4-17. Compare Your Results
in Table 4-16 with These Calculated Values.**

	(All Voltages in Ac Peak)			
Signal Frequency	Input Voltage	(mV)	Output Voltage	Gain
100	0.0050	(5)	2.190	438
100	0.0075	(7.5)	3.285	438
100	0.0100	(10)	4.380	438
500	0.0050	(5)	1.935	387
500	0.0075	(7.5)	2.902	387
500	0.0100	(10)	3.870	387
1000	0.0050	(5)	1.860	372
1000	0.0075	(7.5)	2.790	372
1000	0.0100	(10)	3.720	372
2000	0.0050	(5)	1.810	362
2000	0.0075	(7.5)	2.715	362
2000	0.0100	(10)	3.620	362
4000	0.0050	(5)	1.770	354
4000	0.0075	(7.5)	2.655	354
4000	0.0100	(10)	3.540	354
8000	0.0050	(5)	1.760	352
8000	0.0075	(7.5)	2.640	352
8000	0.0100	(10)	3.520	352
10000	0.0050	(5)	1.755	351
10000	0.0075	(7.5)	2.633	351
10000	0.0100	(10)	3.510	351

ding C4 and R7, as shown in the complete schematic diagram (Fig. 4-26) of the circuit.

We will use a value of 0.7 μF for C_f and a 6.8K (68000 ohm) resistor for R_f. The 0.7 μF capacitance is not a standard value, but it can be achieved by using a 0.47 μF and a 0.22 μF capacitor in parallel.

The formula for the gain including feedback is:

**Table 4-18. A Summary of the Feedback
Impedances at Each of the Test Frequencies.**

C_f = 0.69 μF (0.47 μF = 0.22 μF)
R_f = 6.8K

F	X_{cf}	Z_f
100	2307	7180
500	461	6816
1000	231	6804
2000	115	6801
4000	58	6800
8000	29	6800
10000	23	6800

$$A_{vf} = \frac{A_v}{1 + A_v \, R3/R3 + Z_f}$$

where A_{vf} is the gain with feedback, A_v is the nominal circuit gain (without feedback). Z_f is the impedance of the $C_f - R_f$ combination. R3, of course, is Q1's emitter resistor.

The impedance factor (Z_f) is made of the resistance of R_f and the reactance (X_c) of C_f. These values cannot be added directly, because of differences in their phase. The equation for this impedance is therefore:

$$Z_f = \sqrt{R_f^2 + X_c^2}$$

The reactance for C_f, and the feedback impedance for each of the seven test frequencies are summarized in Table 4-18.

At 100 Hz, Z_f has a value of 7180. Therefore, the modified gain at this frequency is:

$$A_{vf} = \frac{438}{1 + 438 \quad 4.7\big/(4.7 + 7180)}$$

$$= \frac{438}{1 + 438 \quad \dfrac{4.7}{7184.7}} = \frac{438}{1 + 438 \, (0.0006542)}$$

$$= \frac{438}{1 + 0.2865255} = \frac{438}{1.2865255} = 340.45$$

Table 4-19. Table Showing the Effects of Feedback on Gain.

Signal Frequency	Z_f	A_v	(% of maximum	A_{vf}	(% of maximum
100	7180	438	100%	340	100%
500	6816	387	88.36%	305	89.71%
1000	6804	372	84.93%	296	87.06%
2000	6801	362	82.65%	290	85.29%
4000	6800	354	80.82%	284	83.53%
8000	6800	352	80.36%	283	83.23%
10000	6800	351	80.14%	282.5	80.09%

We won't go through the calculation in the test for each individual test frequency. The results are summarized in Table 4-19.

There is still some difference in the gain at different frequencies. Low frequencies receive more gain than high frequencies. But the range of gains is narrowed. Without the feedback network the gain ranged from 351 at 10,000 Hz to 438 at 100 Hz. The highest frequency is subjected to only 80.14% of the gain applied to the lowest frequency.

But, when we add the feedback components, the gain range changes to 282.5 for 10,000 Hz to 340 for 100 Hz. Now the highest frequency's gain is 83.09% of the lowest frequency's gain. This may not seem like a very significant change, but the frequency response of the amplifier is much smoother and stability is considerably improved.

If you repeat the fourth part of the experiment with the modified circuit (including C_f and R_f), you should see an improvement in the reproduction of the waveform at the output when viewed on an oscilloscope.

To further reduce the difference between the low frequency gain and the high frequency gain, a high-pass filter can be added to the circuit to lower the amplitude of low frequency signals.

Actually, in an audio amplifier, this won't always be necessary. Because of the way the human ear responds to different frequencies, a little extra boost in the bass region (low frequencies) will often result in more natural-sounding reproduction.

This completes our sample design of a typical two-stage amplifier circuit. I intentionally used a fairly inefficient design to introduce you to some of the typical problems a circuit designer often faces.

If you want to use this circuit as an audio amplifier driving a loudspeaker, you will have to correct for the impedance mismatch. The output impedance is approximately 50K (50,000 ohms). A typical speaker has an impedance of 8 ohms or so. The easiest solution is to use an impedance matching transformer between the amplifier and the speaker, as illustrated in Fig. 4-32.

Chapter 5

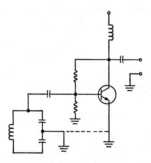

Oscillator Circuits

If the output of an amplifier is fed back to the input in phase, the signal will be reinforced, soon exceeding the limits of the transistor's capabilities. A phantom signal of high amplitude will continue, even if the original input signal is removed. This effect is known as oscillation, and it can be caused by positive feedback in an amplifier.

While highly undesirable in an amplifier circuit, oscillation can be quite useful in other applications. An oscillator is a circuit which generates a predictable, controllable ac signal.

The term oscillator is often used to include waveform generators (see Chapter 4), but strictly speaking, this term refers to a circuit which produces a very pure ac signal known as a *sine wave*. The name comes from the fact that this waveshape looks like a graph of the trigonometric sine function. A sine wave is shown in Fig. 5-1.

The output frequency of most oscillators is determined by some kind of resonance circuit.

REACTANCE AND IMPEDANCE

Before discussing resonance, let's first review the concepts of reactance and impedance. Reactance is ac resistance. Impedance is the combination of reactance and dc resistance. Both reactance and impedance vary with the applied frequency. Dc resistance holds a constant value, regardless of the input signal.

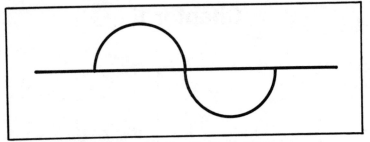
Fig. 5-1. The simplest type of ac signal is the sine wave.

There are two types of reactance. Capacitive reactance is the ac resistance of a pure capacitance. Inductive reactance is the ac resistance of a pure inductance. (Practical capacitors also exhibit some inductive reactance, and practical inductors also exhibit some capacitive reactance, but the effects are generally negligible. In addition, practical capacitors and inductors also exhibit dc resistance.)

Capacitive reactance decreases as the applied frequency increases. The formula is:

$$X_c = \frac{1}{2 \pi FC}$$

where X_c is the capacitive reactance in ohms, π is the mathematical constant pi (approximately 3.14), F is the applied frequency in Hz, and C is the capacitance in farads.

Let's consider an example. We will assume we need to find the capacitive reactance of a 0.33 μF (0.00000033 farad) capacitor at an applied frequency of 2500 Hz. The capacitive reactance in this case words out to:

$$X_c = \frac{1}{2 \times 3.14 \times 2500 \times 0.00000033}$$

$$= \frac{1}{0.0051836} = 193 \text{ ohms}$$

If we increase the applied frequency to 12,000 Hz, the capacitive reactance drops to:

$$X_c = \frac{1}{2 \times 3.14 \times 12000 \times 0.00000033}$$

$$= \frac{1}{0.0248814} = 40 \text{ ohms}$$

Capacitive reactance decreases as the frequency increases.

Now, lets leave the frequency at 12,000 Hz, but increase the capacitance to 1.5 μF (0.0000015 farad). Now the capacitive reactance is:

$$X_c = \frac{1}{2 \times 3.14 \times 12000 \times 0.0000015}$$

$$= \frac{1}{0.1130973} = 8.8 \text{ ohms}$$

Capacitive reactance also decreases as the capacitance increases.

Inductive reactance works in just the opposite way. It increases as the frequency, or the inductance increases. The formula for inductive reactance is:

$$X_L = 2 \pi FL$$

where X_L is the inductive reactance in ohms, π is pi (about 3.14). F is the frequency in Hertz, and L is the inductance in henries.

Again, let's work our way through a few typical examples. We will assume we are working with a 50 mH (0.05 henry) coil. The applied frequency is 2500 Hz. This means the inductive reactance should be equal to:

$$X_L = 2 \times 3.14 \times 2500 \times 0.05 = 785 \text{ ohms}$$

If we increase the applied frequency to 12,000 Hz, the inductive reactance of this coil becomes:

$$X_L = 2 \times 3.14 \times 12000 \times 0.05 = 3770 \text{ ohms}$$

The inductive reactance increases as the applied frequency increases.

Similarly, if we hold the frequency constant (12,000 Hz), and increase the inductance to 120 mH (0.12 henry), the inductive reactance increases to:

$$X_L = 2 \times 3.14 \times 12000 \times 0.12 = 9048 \text{ ohms}$$

Inductive reactance also increases with increases in inductance.

If a circuit contains both capacitive reactance and inductive reactance, one goes up and the other goes down with changes in the applied frequency.

Impedance, as we said earlier, is the combination of dc resistance, capacitive reactance, and inductive reactance. Because of phase differences, these values cannot be simply added together. In a purely resistive circuit, the voltage and the current are in phase. In a pure capacitive inductance, the voltage lags the current by 90°. In a pure inductive reactance, the voltage leads the current by 90°. Capacitive reactance and inductive reactance are 180° out of phase with each other. This implies that they tend to cancel each other out.

The total reactance in a circuit comprised of both capacitive and inductive elements is simply the difference between the inductive reactance and the capacitive reactance:

$$X_T = X_L - X_c$$

If this result is positive, the total reactance is primarily inductive—the voltage leads the current. If the result comes out negative, the total reactance is primarily capacitive—the voltage lags the current.

This total reactance can be combined with the dc resistance to find the impedance, by using this equation:

$$Z = \sqrt{R^2 + X_T^2}$$

Z is the impedance in ohms.

In the following examples, we will use the simple circuit shown in Fig. 5-2. For simplicity, R will be assumed to be a pure resistance,

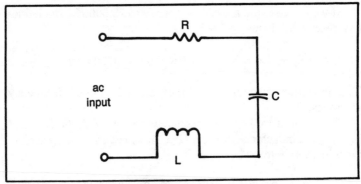

Fig. 5-2. A simple series resonant circuit.

C will be assumed to be a pure capacitance, and L will be assumed to be a pure inductance.

To start, we will assume the following component values:

$$R = 470 \text{ ohms}$$
$$C = 0.33 \text{ } \mu F$$
$$L = 50 \text{ mH}$$

We already know the reactances of the capacitor and inductor for an applied frequency of 2500 Hz:

$$X_c = 193 \qquad X_L = 785$$

The total reactance in this circuit at 2500 Hz is:

$$X_T = X_L - X_c = 785 - 193 = 592$$

The result is positive, so the overall reactance is essentially of the inductive type.

Next, we solve for the impedance of this circuit at 2500 Hz:

$$Z = \sqrt{R^2 + X_T^2} = \sqrt{470^2 + 592^2}$$

$$= \sqrt{220900 + 350464} = \sqrt{571364}$$

$$= 756 \text{ ohms}$$

If we increase the applied frequency to 12,000 Hz, we already know the capacitive reactance will drop to 40 ohms, and the inductive reactance will go up to 3770 ohms. This makes the total reactance in the circuit change to:

$$X_T = 3770 - 40 = 3730$$

Once again, the total circuit reactance is primarily inductive.

The circuit impedance at 12,000 Hz is therefore about equal to:

$$Z = \sqrt{470^2 + 3730^2} = \sqrt{220900 + 13912900}$$

$$= \sqrt{14133800} \cong 3760 \text{ ohms}$$

Next, let's see what happens when we drop the applied frequency to 300 Hz. Now the inductive reactance is:

$$X_L = 2 \times 3.14 \times 300 \times 0.05 = 94 \text{ ohms}$$

And the capacitive reactance now works out to:

$$X_c = \frac{1}{2 \times 3.14 \times 300 \times 0.00000033}$$

$$= \frac{1}{0.000622} = 1608 \text{ ohms}$$

This makes the total reactance in the circuit equal to:

$$X_T = 94 - 1608 = 1514 \text{ ohms}$$

The negative sign indicates that the total circuit reactance is primarily negative. This does not change the impedance equation at all:

$$Z = \sqrt{470^2 + (-1514)^2} = \sqrt{220900 + 2292196}$$

$$= \sqrt{251309} = 1585 \text{ ohms}$$

The reactance sign becomes irrelevant in the impedance equation. It is merely an indication of the phase relationship.

What happens if we hold the applied frequency constant (we'll stick with 300 Hz for now) and change the component values?

First, we will drop the value of C to 0.022 μF (0.000000022 farad). L and R will retain their original values for the time being.

The new capacitive reactance at 300 Hz becomes:

$$X_c = \frac{1}{2 \times 3.14 \times 300 \times 0.000000022}$$

$$= \frac{1}{0.0000415} = 24.114 \text{ ohms}$$

The inductive reactance is still 94 ohms. The total circuit reactance is now:

$$X_T = 94 - 24114 = 24020$$

The capacitive reactance really wipes out the inductive reactance this time.

Now, solving for the impedance, we find it has a value of:

$$Z = \sqrt{470^2 + (-24020)^2} = \sqrt{220900 + 576960400}$$

$$= \sqrt{57718130} = 24025 \text{ ohms}$$

The dc resistance is so small with respect to the reactance, that it has little effect on the impedance. The reverse situation may also occur with some circuits.

For our next example, we hold all values to their levels in the last example, except we will increase the inductance of L to 250 mH (0.25 henry).

At 300 Hz, the inductive reactance of 250 mH is:

$$X_L = 2 \times 3.14 \times 300 \times 0.25 = 471 \text{ ohms}$$

The total reactance in the circuit is now:

$$X_T = 471 - 24114 = 23643 \text{ ohms}$$

So the impedance of the circuit is now equal to:

$$Z = \sqrt{470^2 + (-23643)^2} = \sqrt{220900 + 558991450}$$

$$\sqrt{559212350} = 23648 \text{ ohms}$$

We will try one more example before moving on. We will leave C at 0.022 μF, L at 250 mH, and the frequency at 300 Hz. This means the total circuit reactance also will not change (-23643). But we will change the dc resistance (R) to 33K (33,000 ohms). This makes the impedance work out to:

$$Z = \sqrt{33000^2 + (-23643)^2}$$

$$= \sqrt{1089000000 + 558991450}$$

$$= \sqrt{1647991450} = 40595 \text{ ohms}$$

You can see that the impedance of a circuit is dependent on

many factors—the dc resistance, the inductance, the capacitance, and the applied signal frequency.

RESONANCE

In the preceding section we mentioned that when the inductive reactance is greater than the capacitive reactance, the total reactance is positive (primarily inductive). Similarly, when the inductive reactance is less than the capacitive reactance, the total reactance is negative (primarily capacitive). There is one final special condition we have not dealt with yet.

If the inductive reactance is exactly equal to the capacitive reactance, they will completely cancel each other, and the total effective reactance will be zero:

$$X_T = X_L - X_c = 0$$

Take a look at how this effects the impedance equation:

$$Z = \sqrt{R^2 + X_T^2} = \sqrt{R^2 + O^2} = \sqrt{R^2} = R$$

In this case, the impedance equals the dc resistance. This is the lowest value the impedance can ever take.

This complete canceling of the reactance factors occurs at only a single specific frequency for any combination of capacitive and inductive components. The effect is called *resonance*, and the frequency at which it occurs is called the *resonant frequency*.

The resonant frequency can be found by using the following formula:

$$F_r = \frac{1}{2\pi\sqrt{LC}}$$

F_r is the resonant frequency in Hertz, π is pi (approximately 3.14), L is the inductance in henries, and C is the capacitance in farads.

For example, let's find the resonant frequency for a 0.33 μF (0.00000033 farad) capacitor and a 50 mH (0.05 henry) coil:

$$F_r = \frac{1}{2\times 3.14 \times \sqrt{(0.05 \times 0.00000033)}}$$

$$= \frac{1}{6.28 \times \sqrt{0.0000000165}}$$

214

$$= \frac{1}{6.28 \times 0.0001285} = \frac{1}{0.0008071}$$

$$= 1239 \text{ Hz}$$

Let's prove that the impedance behaves the way we described at the resonant frequency.

First, we calculate the capacitive reactance:

$$X_c = \frac{1}{2 \pi FC} = \frac{1}{2 \times 3.14 \times 1239 \times 0.00000033}$$

$$= \frac{1}{0.002569} = 389.25 \text{ ohms}$$

The inductive reactance at the resonant frequency works out to be:

$$X_L = 2 \pi FL = 2 \times 3.14 \times 1239 \times 0.05$$

$$= 389.24 \text{ ohms}$$

The slight difference between X_c and X_L is due to rounding off fractional values in the various equations. For all intents and purposes the two reactances are equal, so they cancel for a total circuit reactance of $X_T = 0$. The impedance equals the dc resistance in a series LC circuit at resonance:

$$Z = R$$

The resonant formula may be algebraically rearranged to select component values to create a specific resonant frequency:

$$C = - \frac{1}{4 \pi^2 F2 L}$$

or

$$L = \frac{1}{4 \pi^2 F^2 C}$$

For instance, let's say we want a circuit resonant at 1000 Hz,

215

and we have already selected a 200 mH (0.2 henry) coil for L. We would then need a capacitor with a value of:

$$C = \frac{1}{4 \times (3.14)^2 \times (1000)^2 \times 0.2}$$

$$= \frac{1}{4 \times 9.87 \times 1000000 \times 0.2}$$

$$= \frac{1}{7895683.5} = 0.00000013 \text{ farad} = 0.13 \ \mu\text{F}$$

In actually building the circuit, you would probably use a standard 0.15 μF or a 0.12 μF capacitor. Either will probably be close enough.

In critical resonant circuits, a small *trimmer capacitor* is often placed across the main capacitor to allow fine tuning of the resonant frequency. This is shown in Fig. 5-3.

Let's try another example where we know the desired resonant frequency (2200 Hz), and the capacitance 0.05 μF (0.00000005 farad), and need to find the appropriate inductance:

$$L = \frac{1}{4 \times (3.14)^2 \times (2200)^2 \ 0.00000005}$$

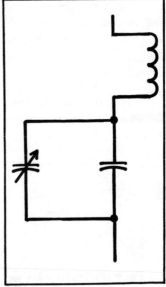

Fig. 5-3. A variable capacitor can be used to make the resonant frequency adjustable.

$$= \frac{2}{4 \times 9.87 \times 4840000 \times 0.00000005}$$

$$= \frac{1}{4.7768885} = 0.2093413 \text{ henry}$$

$$\cong 0.2 \text{ henry} = 200 \text{ mH}$$

If either of these equations gives a really "off-beat," or unavailable value, just select another value for L or C and try again. The same resonant frequency can be reached with many different combinations of components.

For instance, we could have started the last sample problem with a capacitance of 0.22 μF (0.00000022 farad), in which case the necessary inductance would be:

$$L = \frac{1}{4 \times (3.14)^2 \times (2200)^2 \times 0.00000022}$$

$$= \frac{1}{4 \times 9.87 \times 4840000 \times 0.00000022}$$

$$= \frac{1}{42.036619} = 0.0237888 = 0.025 \text{ henry}$$

$$= 25 \text{ mH}$$

This combination gives us the same resonant frequency (2200 Hz) as in the previous example.

Notice that the dc resistance has no effect on the resonant frequency.

PARALLEL LC CIRCUITS

In our discussion so far, we have been assuming that the capacitance and the inductance were in series, as shown in Fig. 5-2. When the capacitance and the inductance are in parallel, as illustrated in Fig. 5-4, we have a somewhat different situation.

The total reactance of a parallel LC circuit works out to:

$$X_T = \frac{(X_L)(X_c)}{(X_L - X_c)}$$

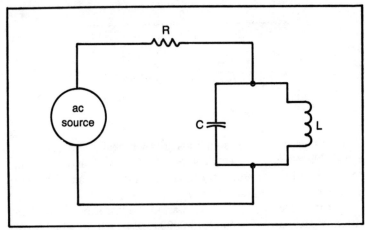

Fig. 5-4. A simple parallel resonant circuit.

The impedance equation is still:

$$Z = \sqrt{R^2 + X_T^2}$$

Let's try an example or two to see how the parallel LC circuit works.

We will start with the same component values we used for our first series LC circuit:

$$R = 470 \text{ ohms}$$
$$C = 0.33 \ \mu F$$
$$L = 50 \text{ mH}$$

We have already solved for these reactances at 2500 Hz:

$$X_c = 193 \qquad X_L = 785$$

For the series LC circuit, the total reactance worked out to 592 ohms, and the impedance at 2500 Hz was 756 ohms.

But now we are working with a parallel LC circuit, so the total reactance is:

$$X_T = \frac{(X_L)(X_c)}{(X_L - X_c)} = \frac{785 \times 193}{785 - 193} = \frac{151505}{592}$$

$$\cong 256 \text{ ohms}$$

A parallel LC circuit behaves differently at resonance than a series LC circuit. At resonance the capacitive reactance equals the inductive reactance, so the total reactance becomes:

$$X_T = \frac{(X_L)(X_c)}{(X_L - X_c)} \quad \frac{(X_L)(X_c)}{0} = \infty$$

Any number divided by zero results in infinity, (or, more correctly, an undefined value).

The impedance becomes:

$$Z = \sqrt{R^2 + \infty^2}$$

By definition, infinity is far larger than any possible value of R, so R can reasonably be eliminated from the equation; leaving:

$$Z = \sqrt{\infty^2} = \infty$$

In a parallel LC circuit, the impedance at resonance is theoretically infinite (with practical components, it won't be true infinity, but it will be extremely high, and we can consider it as infinite in all our equations). Obviously the impedance takes its maximum value at the resonant frequency in a parallel LC circuit.

This is just the opposite of what happened in the series LC circuit, where the impedance reached its minimum value ($Z = R$) at resonance.

OSCILLATION AND RESONANT CIRCUITS

Most oscillator circuits are designed around some sort of parallel LC circuit. The output signal frequency is the resonant frequency of the LC combination.

Consider the circuit shown in Fig. 5-5. This is not a practical circuit, but it is useful for illustrating the concept of parallel resonant oscillation.

When the switch is closed, the voltage through coil L1 will rapidly increase from zero to the source voltage. Of course, the current through the coil must also be increasing during this brief time. A change in the current flow through an inductor will cause a magnetic field to be generated around the coil. This magnetic field cuts through coil L2, inducing a similar voltage in it. This voltage is stored in capacitor C.

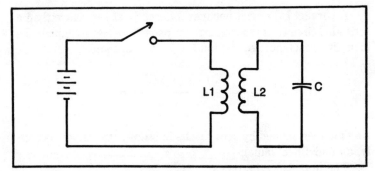

Fig. 5-5. Circuit is used to demonstrate the concept of oscillation.

Soon the voltage across L1 reaches the source voltage. It can't increase beyond that point. The current through this coil becomes constant, and the magnetic field around the coil collapses. No further voltage is induced in L2. C can now discharge its stored voltage back through L2.

This discharge voltage builds up a magnetic field around L2, which induces a voltage in L1, which, in turn induces the voltage back into L2, charging capacitor C in the opposite direction, as shown in Fig. 5-6.

Once the induced voltage in the coils collapses, this whole discharge-induce-charge process repeats with the polarities again reversed, as shown in Fig. 5-7.

If we look at the voltage circulating through the parallel LC network, we will see the voltage build smoothly from zero to a

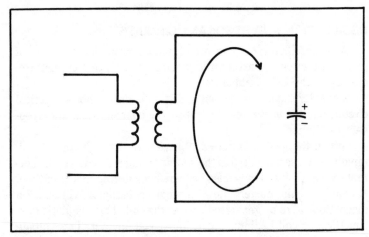

Fig. 5-6. The capacitor begins to charge in the reverse direction.

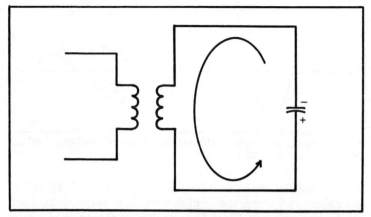

Fig. 5-7. The polarity of the charge on the capacitor continually reverses itself.

positive peak, then smoothly drop back through zero to a negative peak, then back up through zero to the positive peak again. If we graph this changing voltage, we will get a sine wave.

Theoretically the voltage will cycle back and forth between the coil and capacitor indefinitely. In the real world, however, the dc resistance of the components will decrease the amplitude on each cycle. This effect is called *damping*. A damped sine wave is illustrated in Fig. 5-8. If no additional energy is introduced into the circuit, the oscillations will quickly drop to zero.

In addition to attenuation due to dc resistance within the circuit, if we tap off any output signal, we will also be consuming some of the energy within the circuit. Obviously, to be of any practical value, an oscillator must have some kind of output.

Clearly, this simplified circuit will not be able to sustain oscillations for very long. That's why this circuit is not at all practical.

Practical oscillator circuits always include some kind of amplifier stage. The output of the oscillator amplifier is fed back to the input in phase (positive feedback). This allows the amplitude of the oscillations to be sustained at a usable level.

It might at first seem that the amplifier would keep increasing the output amplitude infinitely, since the output signal is continuously being passed back through the amplifier for more amplification. But all amplifiers have natural limitations that will prevent any further increases beyond some specific output level. No amplifier can put out more power than it receives from its power supply.

The amplification also maintains the energy within the parallel LC circuit, or tank, at a constant level, so the oscillations will not

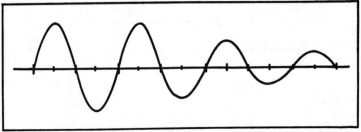

Fig. 5-8. If no outside energy is added to the LC tank, the oscillations will die out, producing a damped sine wave.

die out. Oscillations will continue as long as power is applied to the circuit. In a sense, the amplification stage keeps opening and closing the switch of Fig. 5-5 many times a second.

There are a number of approaches to creating an oscillator circuit, but they are all essentially variations on the same theme.

THE HARTLEY OSCILLATOR

One of the most common forms of transistor oscillator circuits is the Hartley oscillator. A simplified version of this circuit is shown in Fig. 5-9.

This circuit is also sometimes called a *split inductance oscillator*. This is because we tap off the oscillations in the center of the coil in the LC tank. Essentially, we split the inductance into two parts.

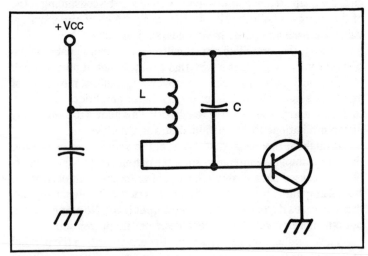

Fig. 5-9. The Hartley Oscillator.

A more practical form of this circuit is shown in Fig. 5-10.

When power is first applied to this circuit, R2 places a small voltage on the base of the transistor, allowing it to start conduction.

If the transistor was perfect and absolutely noise free, nothing much would happen. However, the oscillator circuit takes advantage of the imperfection of real-world components. The transistor amplifier will generate some small amount of internal noise. This random signal will be fed repeatedly back to the input for positive feedback and re-amplification. It will quickly reach a usable level.

When this noise signal reaches a usable level, current from the collector will pass through R4, R2, and R1, finally reaching the parallel resonant tank (L1 and C1).

The current enters the tank through the AB section of the coil. The rising current through this section of the coil will induce a voltage in coil section BC. This voltage is stored by C1.

C2 is selected so it has an extremely low impedance at the

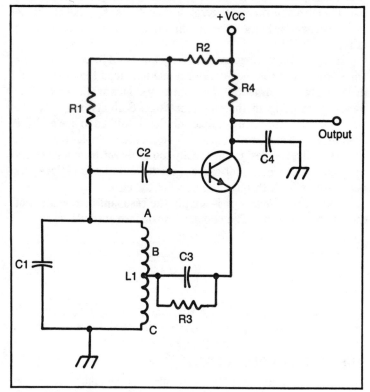

Fig. 5-10. A more practical Hartley Oscillator circuit.

oscillating frequency (the resonant frequency of L2 and C1). As a result of this low impedance, the base of the transistor is connected more or less directly to C1. (The initiating base voltage supplied through R2 is quite low, so it can be reasonably ignored once oscillations begin. It is only needed to get the oscillations started. After that, it serves no real purpose.)

As C1 charges, it increases the bias on the transistor. This, in turn, will increase the current through coil section AB, and the induced voltage through coil section BC. At the same time, the charge on both C1 and C2 is steadily increased.

Eventually the voltage from C1 will equal the R1-C2 voltage, but with the opposite polarity. In other words, these voltages cancel out. At this point, the transistor is saturated. Its output current stops rising, so the magnetic field around coil section AB collapses. C1 starts to discharge through L1. As the current through the coil increases, it builds up a magnetic field. Once the capacitor is fully discharged, the coil will tend to oppose the change in current flow (i.e., the current dropping to zero), so it continues to conduct, charging the capacitor in the opposite direction, and the entire process is repeated.

In some applications the relatively low input impedance of the transistor in the common-emitter mode may load the tank circuit excessively (i.e., draw too much current). This will decrease the power circulating in the resonant tank. Stability may also be adversely affected. One possible solution would be to use an FET (Field-Effect Transistor) in place of the standard bipolar transistor.

Basically, the design isn't really too different from a standard common-emitter amplifier stage, except the input is replaced by positive feedback through the parallel LC tank.

The output frequency is simply the resonant combination of the LC combination. The standard resonance formula is:

$$F_r = \frac{1}{2\pi\sqrt{LC}}$$

Resistor R2 should have a very high value so only a small dc voltage is placed on the base of the transistor to initiate oscillations.

THE COLPITTS OSCILLATOR

Closely related to the Hartley oscillator is the Colpitts oscillator, which is shown in its basic form in Fig. 5-11. The primary difference

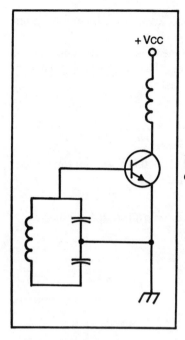

Fig. 5-11. The basic Colpitts Oscillator circuit.

is that instead of a split inductance in the resonant tank (as with the Hartley oscillator), the Colpitts oscillator features a split capacitance.

A practical variation of the basic Colpitts oscillator is shown in Fig. 5-12. Resistors R1 and R2 put the dc bias on the base. They should each have a value of at least 1000 ohms. Coil L2 is a RF choke. Capacitor C3 blocks the dc bias voltage from the parallel resonant tank (L1, C1, and C2). It serves the same basic function as C2 in Fig. 5-10. Finally, C4 simply blocks the dc VCC supply from the output. The ac oscillations pass through this capacitor.

The basic operation of this circuit is essentially the same as the operation of the Hartley oscillator described in the preceding section.

The parallel LC circuit is made of coil L1 and capacitors C1 and C2. These two capacitors in series function as a single capacitor as far as the LC resonant circuit is concerned. The center-tap between them provides a feedback path back to the transistor's emitter through the circuit ground.

If the two capacitors are of equal value, the total effective capacitance within the LC network (determining the resonant frequency) will be equal to one-half the value of either capacitor separately. In most designs C1 and C2 will probably not be given equal values.

If the two capacitances within the tank are of unequal values, the total effective tank capacitance may be calculated using the standard formula for capacitors in series:

$$\frac{1}{C_T} = \frac{1}{C1} + \frac{1}{C2}$$

For example, if C1 is 0.01 μF and C2 is 0.005 μF, the total effective capacitance would work out to:

$$\frac{1}{C_T} = \frac{1}{0.01} + \frac{1}{0.005} = 100 + 200 = 300$$

$$C_T = \frac{1}{300} = 0.0033 \ \mu F$$

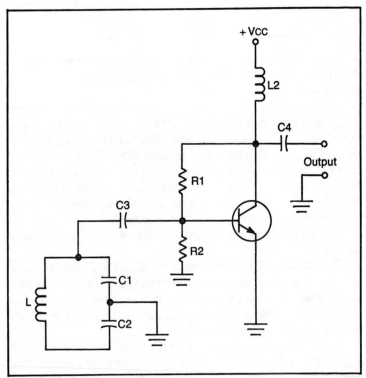

Fig. 5-12. A practical Colpitts Oscillator circuit.

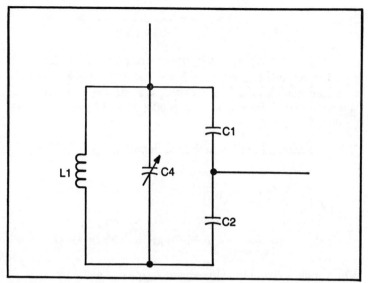

Fig. 5-13. Placing a variable capacitor in parallel with the LC tank makes the output frequency manually adjustable.

In actual practice, the two capacitors are usually of unequal values, because the strength of the feedback signal is dependent on the ratio between these two capacitances. By changing both of these capacitors in an inverse fashion, the feedback level can be varied, while the resonant frequency remains constant. When C1 is increased, C2 would be decreased by a similar amount, or vice versa.

This points up one of the chief drawbacks of the Colpitts oscillator. When the frequency is changed, we don't want the feedback signal to vary or the level of the output signal will not be constant. In fact, in some cases, oscillation may not be sustained when one of the tank capacitances is changed. This means that we have to change both capacitances simultaneously.

True, we could make the coil adjustable instead of the capacitors, but in most applications this would probably be just as impractical. Generally speaking, it is preferable and more convenient to use an adjustable capacitor, rather than an adjustable coil.

A common solution to this problem with the Colpitts oscillator is illustrated in Fig. 5-13. This method keeps the C1:C2 ratio constant, while the resonant frequency is readily adjustable. The total effective capacitance in the tank circuit is determined by the following formula:

$$C_T = C_x + \frac{1}{1/C1 + 1/C2}$$

Let's see how this works by trying a couple of examples. Let's say L1 is 100 mH (0.1 henry), C1 is 0.1 μF (0.0000001 farad), C2 is 0.22 μF (0.00000022 farad), and C_x is a variable capacitor with values that range from 0.0001 μF (0.0000000001 farad) to 0.01 μF (0.00000001 farad).

First, the nominal total capacitance (ignoring C_x) is:

$$\frac{1}{C_T} = \frac{1}{0.1} + \frac{1}{0.22} = 10 + 4.545 = 14.545$$

$$C_T = \frac{1}{14.545} = 0.06875 \ \mu F = 0.00000006875 \ farad$$

Which makes the nominal resonant frequency equal to:

$$F_R = \frac{1}{2 \times 3.14 \times \sqrt{(0.1 \times 0.00000006875)}}$$

$$= \frac{1}{2 \times 3.14 \times \sqrt{0.000000006875}}$$

$$= \frac{1}{6.28 \times 0.0000829} = \frac{1}{0.000521}$$

$$= 1919.5 \ Hz$$

When we add the effects of C_x, the minimum total tank capacitance becomes:

$$C_T = 0.0001 + \frac{1}{1/0.1 + 1/0.22}$$

$$= 0.0001 + \frac{1}{14.545} = 0.0001 + 0.06875$$

$$= 0.06855 \ \mu F = 0.00000006885 \ farad$$

So the resonant frequency becomes:

$$F_R = 2 \times 3.14 \times \dfrac{1}{\sqrt{0.1 \times 0.00000006885}}$$

$$= 2 \times 3.14 \times \dfrac{1}{\sqrt{0.000000006885}}$$

$$= \dfrac{1}{6.28 \times 0.000083} = \dfrac{1}{0.0005214}$$

$$= 1918 \text{ Hz}$$

Because C_x is so small compared to C1 and C2, it doesn't have too much of an effect on the resonant frequency (only 1.5 Hz in this example).

Now, if we raise C_x to its maximum value (0.01 μF), the total tank capacitance is increased to:

$$C_T = 0.01 + 0.06875 = 0.07875 \ \mu\text{F}$$

$$= 0.00000007875 \text{ farad}$$

This changes the resonant frequency to:

$$F_R = 2 \times 3.14 \times \dfrac{1}{\sqrt{0.1 \times 0.00000007875}}$$

$$= 6.28 \times \dfrac{1}{\sqrt{0.000000007875}}$$

$$= \dfrac{1}{6.28 \times 0.0000887} = \dfrac{1}{0.0005576}$$

$$= 1793.5 \text{ Hz}$$

Notice that C_x always increases the capacitance, so the resonant frequency is always less than its nominal (no C_x) value.

DESIGNING A COLPITTS OSCILLATOR

We will now work through the step-by-step design procedure for a typical Colpitts oscillator. We will aim for the following specifications:

> VCC = 9 volts
> output frequency = 10 kHz (10000 Hz)
> output power = 35 mW (0.035 watt)

The amplifier stage will be operated in class A for maximum stability.

We will design our oscillator using the standard circuit diagram shown in Fig. 5-14. We will use a HEP-50 general purpose NPN transistor. A partial specification sheet for this device is given in Table 5-1.

The load resistance of the LC tank should be equal to:

$$R_L = \frac{VCC^2}{2P_o}$$

where P_o is the output power.

It is a good idea to cut back on the VCC voltage a little, to leave some margin for error. Since we are using a 9-volt power supply or battery, we can lower VCC to 8 volts. This makes the load resistance equal to:

Table 5-1. The HEP-50 Transistor.

TYPE	NPN
β	85
ABSOLUTE MAXIMUM RATINGS	
POWER	400 mW (0.4 watt)
I_c (Collector current)	300 mA (0.3 ampere)
V_{cb}	25 volts
V_{ce}	15 volts
V_{eb}	4.0 volts
FREQUENCY RESPONSE 250 MHz (250,000,000 Hz)	

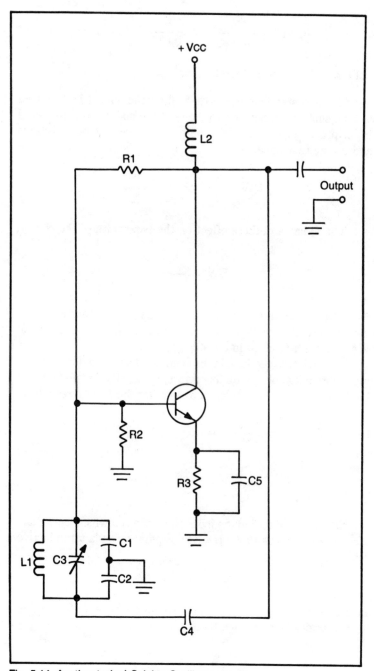

Fig. 5-14. Another typical Colpitts Oscillator circuit.

$$R_L = \frac{8^2}{2 \times 0.035} = \frac{64}{0.07} = 914 \text{ ohms}$$

Let's round this value to 900 ohms.

The next step is to select the coil for the tank (L1). We need to give some consideration to the Q of the coil. Q was discussed in the preceding chapter. It is the ratio between the dc resistance and the ac reactance:

$$Q = \frac{X}{R_s} \quad \text{or} \quad Q = \frac{R_p}{X}$$

The Q has a definite effect on the bandwidth of the circuit:

$$BW = \frac{F_o}{Q}$$

where F_o is the output frequency, and BW is the bandwidth.

For a typical oscillator, we usually want a Q of about 10 to 20. We will use a Q of 20 in our design.

We are interested here in the Q imposed on the coil by paralleling it with a 900-ohm load. Therefore, we use the parallel form of the Q equation, and rearrange it to solve for the desired reactance:

$$Q = \frac{R_p}{X} = \frac{900}{X} = 20$$

$$X = \frac{R_p}{Q} = \frac{900}{20} = 45 \text{ ohms}$$

Now we need to find the inductance of a coil which has a reactance of 45 ohms at 10000 Hz. The general formula for inductive reactance is:

$$X_L = 2 \pi FL$$

In this case, we know $X_L = 45$, but we do not know L. Therefore, we algebraically rearrange the equation:

$$L = \frac{X_L}{2 \pi F} = \frac{45}{2 \times 3.14 \times 10000}$$

$$= \frac{45}{62832} = 0.0007162 \text{ henry} = 0.72 \text{ mH} = 720 \text{ } \mu\text{H}$$

If we round that off to 750 μH, the actual inductive reactance at 10000 Hz should be:

$$X_L = 2 \pi \text{ FL} = 2 \times 3.14 \times 10000 \times 0.00075$$

$$= 47 \text{ ohms}$$

which changes the Q to:

$$Z = \frac{R_p}{X} = \frac{900}{47} = 19.1$$

That is certainly close enough.

At resonance, the capacitive reactance should be equal to the inductive reactance. The formula for finding the capacitive reactance, you should recall, is:

$$X_c = \frac{1}{2 \pi \text{ FC}}$$

This equation can easily be rearranged to solve for the capacitance from a known capacitive inductance:

$$C = \frac{1}{2 \pi \text{ FX}_c} = \frac{1}{2 \times 3.14 \times 10000 \times 45}$$

$$= \frac{1}{2827433} = 0.00000035 \text{ farad} = 0.35 \text{ } \mu\text{F}$$

This is the total estimated capacitance in the LC tank, i.e., the parallel combination of C1 and C2 (we will ignore the effects of tuning capacitor C3 for now). We will temporarily assign equal values to the two capacitors. This means each will take a value of 0.175 μF. This is not a standard capacitor value, of course, but we don't have to worry about that for now, since these are only temporary values for our calculations.

Now, let's consider the effects of fine-tuning capacitor C3. This capacitor is selected to give us a range of user-selectable output frequencies. If a narrow range is desired, especially if we only need C3 to fine-tune the output frequency, only a small capacitance is needed. A larger variable capacitor gives a greater range.

Let's assume that C3 is a 365 pF (0.000000000365 farad) tuning capacitor. This value was selected because it is widely available. It is the size often used to tune portable AM radios. We will further assume that at its minimum setting its capacitance is 5 pF (0.000000000005 farad).

We already know the series combination of C1 and C2 is 0.35 μF (0.00000035 farad). We will call this C_s. We can simplify the equation for the parallel combination to:

$$C_T = C3 + C_s$$

When C3 is at its minimum setting, its effect on the total capacitance is negligible:

$$C_T = 0.000005 + 0.35 = 0.350005 \ \mu F$$
$$= 0.\ 000000350005 \text{ farad}$$

The output frequency (resonance) at this setting is:

$$F_R = \frac{1}{2^\pi \sqrt{LC_T}}$$

$$F_R = \frac{1}{2 \times 3.14 \times \sqrt{0.00075 \times 0.000000350005}}$$

$$= \frac{1}{6.28 \times \sqrt{0.00000000026250375}}$$

$$= \frac{1}{6.28 \times 0.0000162} = \frac{1}{0.0001018}$$

$$= 9823 \text{ Hz}$$

That's a little lower than our target frequency of 10,000 Hz. If this was a critical application we would need to go back and select new values for L1, C1 and C2 to correct for the cumulative rounding errors. For now, we will assume our application isn't terribly critical, so 9823 Hz will be close enough. This is the maximum frequency our circuit will be capable of.

Now, let's consider what happens when C3 is at its maximum setting. The total tank capacitance is now:

$$C_T = 0.000365 + 0.35 = 0.350365 \ \mu F$$
$$= 0.000000350365 \text{ farad}$$

This makes the resonant frequency equal to:

$$F_R = \frac{1}{2 \times 3.14 \times \sqrt{0.00075 \times 0.000000350365}}$$

$$= \frac{1}{6.28 \times \sqrt{0.0000000002652375}}$$

$$= \frac{1}{6.28 \times 0.0000163} = \frac{1}{0.0001023}$$

$$= 9772.5 \text{ Hz}$$

Because C3's maximum value is relatively small with respect to C_s (the series combination of C1 and C2), the range of output frequencies is rather small. The output frequency with the component values listed in this experiment can range from 9772.5 Hz to 9823 Hz—a range of 50.5 Hz. This is suitable for fine-tuning. If we want an oscillator that can cover a wider range of output frequencies, we need to increase the tuning range of C3.

Let's replace C3 with a tunable capacitor with a maximum value of 0.05 μF. For convenience, we will assume the minimum setting is still 5 pF. This way, the maximum frequency remains 9823 Hz. We don't have to redo those calculations.

When C3 is set to its maximum value, the total tank capacitance is:

$$C_T = 0.05 + 0.35 = 0.4 \ \mu F = 0.0000004 \text{ farad}$$

This gives us a minimum resonant frequency of:

$$F_R = \frac{1}{2 \times 3.14 \times \sqrt{0.00075 \times 0.0000004}}$$

$$= \frac{1}{6.28 \times \sqrt{0.0000000003}}$$

$$= \frac{1}{6.28 \times 0.0000173} = \frac{1}{0.0001088}$$

$$= 9189 \text{ Hz.}$$

Now we have a total output range of:

$$\text{Range} = 9823 - 9189 = 634 \text{ Hz}$$

For large output ranges, select a large value for tuning capacitor C3. For small, fine-tuning output ranges, use a relatively small value for tuning capacitor C3.

Capacitors C1 and C2 function as a voltage divider. If they are equal, half of the power within the LC trap will reach the transistor's emitter through ground.

In a practical design we want to limit the amount of feedback. If there is too much feedback, the repeated reamplification will cause the transistor to put out a voltage greater than its source voltage. Clearly that is impossible. The signal will be clipped and badly distorted, as illustrated in Fig. 5-15. In addition, some transistors may build up excessive heat if overdriven like this for an extended period of time. This excess heat could damage the delicate semiconductor crystal.

The feedback could be limited somewhat by placing a resistor (R3) between the emitter and ground. A capacitor (C5) in parallel with the emitter resistor will improve circuit stability.

The voltage gain is equal to:

$$A_v = \beta \; \frac{R_L}{Z_i}$$

where R_L is the load resistance, A_v is the voltage gain, Z_i is the input impedance of the transistor, and β, of course, is the transistor's beta.

We already have determined that $R_L = 900$ ohms. We can get the beta from the specification sheet for the transistor:

$$\beta = 85$$

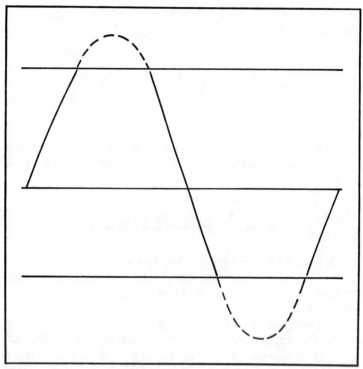

Fig. 5-15. If the oscillations are too strong, the waveform will be clipped and distorted.

The formula for the input impedance is:

$$Z_i = \beta\left(\frac{26}{I_e}\right) \cong \beta\left(\frac{26}{I_c}\right)$$

The maximum current drawn by the amplifier stage is:

$$I_c = \frac{VCC}{R_L} = \frac{9}{900} = 0.01 \text{ ampere} = 10 \text{ mA}$$

Since the quiescent operating point should be half the maximum range, we can assign a value of 5 mA to I_c. This makes the input impedance equal to:

$$Z_i = 85 \times \left(\frac{26}{5}\right) = 85 \times 5.2 = 442 \text{ ohms}$$

Now we have enough information to find the voltage gain of the amplifier stage:

$$A_v = \beta \left(\frac{R_L}{Z_i} \right) = 85 \times \frac{900}{442}$$

$$= 85 \times 2.04 \cong 173$$

A good rule of thumb in designing oscillators is to divide this value by about 4, and round off to a convenient value. This gives us a voltage gain of:

$$A_{va} = \frac{A_v}{4} = \frac{173}{4} = 43.25 \cong 45$$

If C1 and C2 are of equal value, they will divide this gain factor in half, giving us an effective gain of 22.5. For maximum stability, however, we want the gain to be slightly better than unity (a little greater than 1).

For best stability (and best power output) we also want matched impedances. The capacitive divider is used to achieve both these goals. Once we have adjusted the C1/C2 ratio for best impedance matching, we can trim the gain by adjusting the value of emitter resistor R3.

The desired value for C2 can be calculated with this formula:

$$C2 = C_T \sqrt{\frac{R_o}{R_i}}$$

where C_T is the total value of the C1/C2 combination in series (0.35 μF for our example). R_o is the output impedance/resistance. It is the same as R_L (900 ohms). R_i is the input impedance/resistance, and has the same value as Z_i (442). So, for our sample design, C2 should be equal to:

$$C2 = 0.35 \times \sqrt{\frac{900}{442}} = 0.35 \times \sqrt{2.04}$$

$$= 0.35 \times 1.43 = 0.5 \ \mu F$$

Now, we just have to rearrange the series combination equation to find the necessary value for C1:

$$C_T = \frac{C1 \times C2}{C1 + C2}$$

$$C1 = \frac{C_T}{1 - C_T/C2} = \frac{0.35}{1 - 0.35/0.5}$$

$$= \frac{0.35}{1 - 0.7} = \frac{0.35}{0.3} = 1.17 \ \mu F = 1.2 \ \mu F$$

We have done most of the design now. Now there are just a few component values we need to select.

First, the emitter resistor (R3) is selected to give the desired gain. One method would be to breadboard the circuit using a trimpot in place of R3. Monitor the oscillator's output with an oscilloscope and adjust R3 for the cleanest possible waveform. Then replace the trimpot with a fixed resistor whose value is approximately equal to the setting of the trimpot.

We can also approximate the value for this resistor by using the gain equation we used for designing amplifiers in Chapter 4:

$$A_v = \frac{R_{cc}}{R_e}$$

The problem with this approach is that R_{cc} is made of the reactances of several components. This makes the calculations very complex. The experimental method is usually more convenient.

If you do not have an oscilloscope available, R3 can be adjusted by running the output through an amplifier and speaker, and adjusting the trimpot for the strongest, purest tone. There should be no raspiness in the tone at all.

This resistor will usually be quite small—generally well under 500 ohms. Values of 50 to 200 ohms will be pretty much the norm.

Capacitor C4 should be selected so its reactance at the resonant frequency is about one tenth the load resistance (R_L):

$$X_{c4} = \frac{R_L}{10} = \frac{900}{10} = 90$$

$$C4 = \frac{1}{2 \pi FX_{c4}} = \frac{1}{2 \times 3.14 \times 10000 \times 90}$$

$$= \frac{1}{5654867} = 0.00000018 \text{ farad} = 0.18 \ \mu F$$

We can round this value to a standard 0.2 μF or 0.22 μF capacitor. These equations are only approximations. The exact values are not very critical.

Coil L2 serves as a choke. To prevent its dc resistance from limiting the output of the oscillator, this coil should have a high reactance compared to the load resistance (R_L). As a rule of thumb, the inductive reactance of L2 at the output frequency should be at least 50 times greater than the load resistance:

$$X_{L2} = 50 \times R_L = 50 \times 900 = 45000 \text{ ohms}$$

This equation gives us a minimum value. Let's boost it up a little, and round it off to 50000 ohms.

Again, we can find the inductance from the inductive reactance by rearranging the original formula for the inductive reactance:

$$L = \frac{X_L}{2 \pi F} = \frac{50000}{2 \times 3.14 \times 10000}$$

$$= \frac{50000}{62832} = 0.796 \text{ henry} \cong 800 \text{ mH}$$

Capacitor C5 will normally have a value equal to that of C4, which is 0.2 μF in our sample design.

Resistor R2 should have a value approximately 5 times that of R3:

$$R2 = 5R3$$

Finally, the formula for R1 is:

$$R1 = \frac{R2VCC}{V_e} - R2$$

where V_e is the voltage drop across R3. Ohm's Law allows us to find that value:

$$V_e = I_e R3$$

And, since the emitter current is roughly equal to the collector current, this can be rewritten as:

$$V_e + I_c R3$$

which is handy, since I_c was found in one of the earlier equations.

Assuming a value of 100 ohms for R3, the other two resistances work out to:

$$R2 = 5 \times 100 = 500 \text{ ohms} \cong 470 \text{ ohms}$$

$$V_e = 0.005 \times 100 = 0.5 \text{ volt}$$

$$R1 = \frac{470 \times 9}{0.5} - 470 = \frac{4230}{0.5} - 470$$

$$= 8460 - 470 = 7990 \cong 8200 \text{ ohms (8.2K)}$$

All our calculated parts values for this sample design are listed in Table 5-2.

VARIATIONS ON THE BASIC COLPITTS OSCILLATOR

The Colpitts oscillator is a very popular circuit, because it offers fairly good frequency stability at a reasonable cost. As with most other frequently used circuits, a number of variations on the basic Colpitts oscillator are often encountered.

**Table 5-2. Calculated Component Values
for the Sample Design of a Colpitts Oscillator.**

Q	HEP50 NPN Transistor
R1	8.2K ohms—see text
R2	470 ohms—see text
R3	100 ohms—see text
C1	1.2 μF
C2	0.5 μF
C3	Adjustable capacitor—see text
C4	0.2 μF
C5	0.2 μF
ΣL1	720 mH
L2	800 mH choke

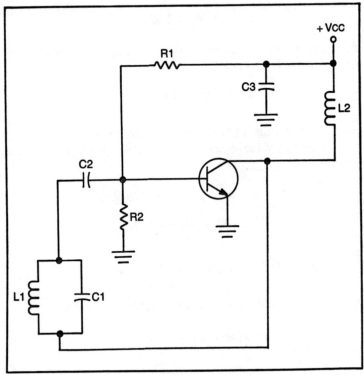

Fig. 5-16. The Ultra-Audion Oscillator.

One fairly common variation of the Colpitts oscillator circuit is the *Ultra-Audion oscillator*, which is shown in Fig. 5-16. This circuit is very well suited to VHF (Very High Frequency) applications.

The feedback paths in this oscillator are provided by the internal capacitances within the transistor itself. Ordinarily these internal capacitances are a problem at high frequencies. In the Ultra-Audion oscillator they are turned into an advantage.

These internal capacitances give the effect of small capacitors between the base and the emitter, and between the emitter and the collector, as shown in Fig. 5-17. These capacitances are quite small, so in the VHF range they exhibit a very low reactance.

The same idea can be used at lower frequencies, but larger external capacitors will have to be used to get capacitive reactances that are low enough at the operating frequency. This is illustrated in Fig. 5-18.

Another variation on the basic Colpitts oscillator circuit is shown in Fig. 5-19. This is the *Clapp oscillator*.

242

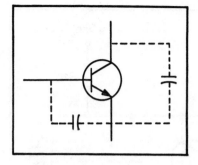

Fig. 5-17. Stray capacitances within the transistor become significant at high frequencies.

The Clapp oscillator is fairly unique in that it is tuned by a series resonant LC network (L1 and C1) rather than the more commonly employed parallel resonant tank.

Feedback for the Clapp oscillator is provided by the two capacitors in the emitter circuit (C3 and C4). These capacitors typically have very small values.

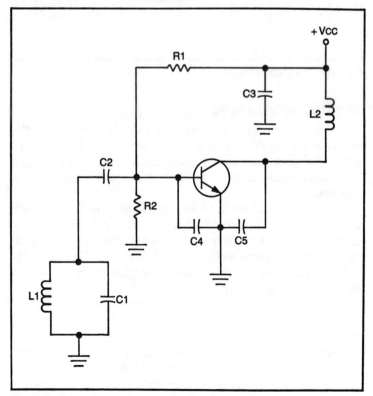

Fig. 5-18. External capacitors used at low frequencies.

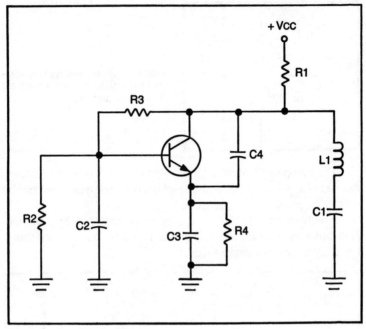

Fig. 5-19. The Clapp oscillator.

CRYSTAL OSCILLATORS

While the frequency stability of LC oscillators like the Hartley oscillator and the Colpitts oscillator can be quite good, it often isn't precise enough for certain critical applications, such as in test equipment, radio transmitters, or the local color oscillator in a television receiver.

In such critical applications, crystal oscillators are usually used. Oscillators built around quartz crystals can be extremely accurate and stable, especially if a constant temperature is maintained. Broadcast stations (both radio and television) keep their carrier frequency oscillators in special "crystal ovens" to hold a precisely constant temperature.

Figure 5-20 shows the electrical equivalent of a crystal. Depending on how the crystal is manufactured (and the relative sizes of capacitances c_a and c_b), it can replace either a series resonant or a parallel resonant LC network. Generally, a crystal designed for series resonant use cannot be used in a parallel resonant circuit.

The resonant frequency of a crystal is determined primarily by the size and thickness of the quartz slab used as the crystal body.

244

Crystals can also be made to resonate at integer multiples of their main resonant frequency. These multiples are called *overtones* or *harmonics*. The primary resonant frequency is called the *fundamental*.

For example, a crystal designed to resonate at 15000 Hz (the fundamental) will also resonate (but to a lesser degree) at 30000 Hz (second harmonic), 45000 Hz (third harmonic), 60000 Hz (fourth harmonic), and so forth. Notice that no "first harmonic" is mentioned, since one times the fundamental is the same as the fundamental frequency. The resonance effect will become steadily less pronounced at higher harmonics.

Crystals are generally more expensive than separate capacitors and coils, but their resonant frequency is much more precise and stable. Capacitor-coil resonant circuits often drift off frequency (that is, the components change their values slightly), particularly under changing temperature conditions. Crystals can also drift off frequency with changes in ambient temperature, but they are considerably less sensitive than coils and capacitors.

Another major advantage of crystals is reliability. Their failure rate tends to be somewhat lower than for capacitors and coils.

However, crystals are fairly delicate devices. They can be damaged by high overvoltages, or extremely high temperatures.

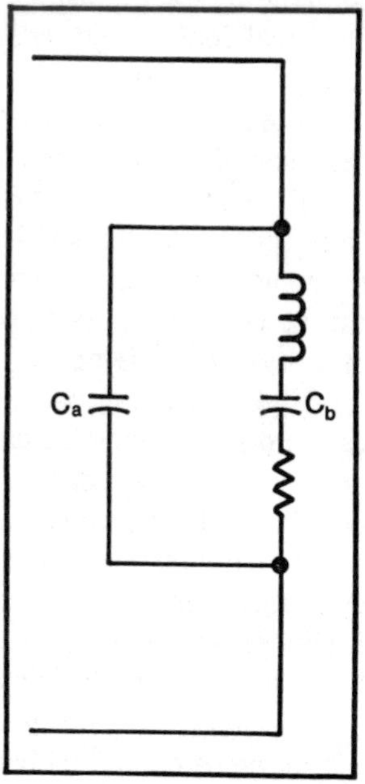

Fig. 5-20. A crystal RLC circuit equivalent.

In addition, a severe mechanical shock (such as being dropped on-to a hard surface) could crack the delicate crystal slab. If this happens, the entire crystal package must be replaced. There is no way to repair or replace just the crystal slice.

Occasionally, the hermetic seal of the crystal's housing can develop a small leak. This could allow contaminants (like dust or water) to get inside the metal case. These contaminants could interfere with the electrical contact between the crystal slab and the device leads.

All these defects are relatively rare, but they do occur frequently enough to warrant mentioning them here. They are also all unrepairable. If any of these problems show up, a new crystal unit must be used.

The most important disadvantage of crystals is that the resonant frequency generally cannot be changed. In an LC circuit, one or both components can be made variable, but there is no such thing as a variable crystal. Sometimes special external circuitry can be added to provide a limited degree of fine-tuning, but generally a crystal resonant circuit is of a fixed frequency. The only way the frequency can be changed is by physically replacing the crystal.

For this reason crystals are usually inserted into sockets, rather than actually being permanently soldered into the circuit. The leads from the crystal's housing plug into holes in the socket which make electrical contact with the external circuitry. Crystals can easily and quickly be removed and replaced. Using sockets also sidesteps the potential problem of thermal damage resulting from soldering the crystal's leads directly.

The size and thickness of the crystal determines the resonant frequency. The thinner the crystal slab, the higher the resonant frequency. Obviously there is a physical limit to how thin a crystal can be cut. This puts a theoretical upper limit on the frequency of a crystal oscillator.

Higher than normal frequencies can be obtained from crystal oscillators by forcing the crystal to oscillate at one of its harmonics, or by passing the signal through special circuits called frequency doublers, which do just what their name implies. For example, if a 5 MHz (5,000,000 Hz) oscillator signal is fed through a frequency doubler, the output will be 10 MHz (10,000,000 Hz). If this signal is now passed through a second frequency doubler, the signal will be raised to 20 MHz (20,000,000 Hz) at the output.

Crystal oscillators generally aren't used for low frequency operation. A crystal with a resonant frequency in the audio range (below 20 kHz (20,000 Hz)) would be very large and bulky. Crystals are commonly used only for RF (Radio Frequency) oscillator circuits.

If the precision of a crystal oscillator is required at a lower frequency, frequency divider circuits could be used. These are plainly just the opposite of the frequency multipliers mentioned earlier. A frequency divider divides the signal frequency by two. For instance, if a 5 MHz (5,000,000 Hz) signal is fed through a frequency divider, the output will be 2.5 MHz (2,500,000 Hz). A second frequency divider stage would drop the signal frequency to 1.25 MHz (1,250,000 Hz).

In a crystal oscillator circuit, an ac signal is applied between the two leads from the crystal. This causes the crystal slab to vibrate, due to the piezoelectric effect. (An electrical stress across the X axis of a crystal will produce a mechanical stress across the Y axis.) Depending on the thickness of the crystal slab, it will be mechanically resonant at some specific frequency—the resonant frequency.

If the applied ac voltage is equal to the crystal's resonant frequency (or an exact harmonic of the crystal's resonant frequency)

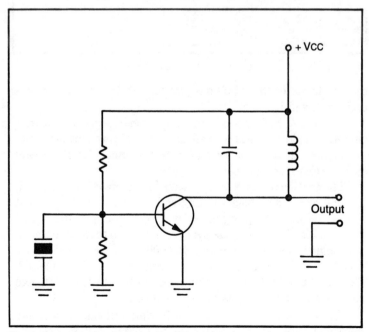

Fig. 5-21. A typical crystal oscillator circuit.

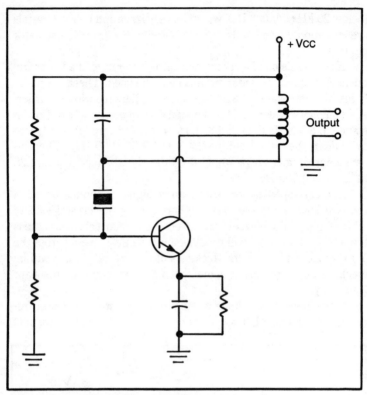

Fig. 5-22. Crystal oscillator in the series resonant mode.

the amplitude of the vibrations will tend to be quite large, and oscillation will be sustained.

Figure 5-21 shows the schematic diagram for a crystal oscillator circuit. Here the crystal is used as a parallel resonant network.

Figure 5-22 shows a crystal oscillator circuit using the crystal in the series resonant mode.

The parallel resonant form is somewhat more commonly used.

One nice thing about working with crystal oscillators is that general purpose designs can be used. For example, the circuit shown in Fig. 5-23 can be used with crystals with resonant frequencies ranging from about 100 kHz (100,000 Hz) to well over 15 MHz (15,000,000 Hz) using the component values listed in Table 5-3. The only thing in the circuit that needs to be changed to change the operating frequency is the crystal itself.

The component values are not terribly critical. Almost any general purpose NPN transistor can be used for Q1. In fact, this

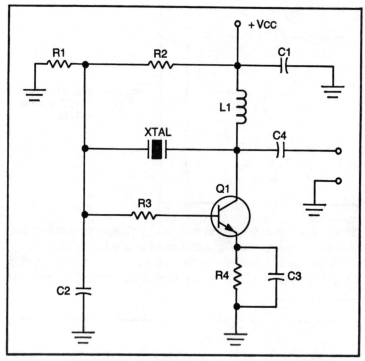

Fig. 5-23. This circuit will work with crystals at a variety of frequencies.

circuit could easily be adapted as a simple GO/NO GO transistor tester.

By using a crystal socket, this circuit can be put to work in a number of applications, simply by plugging in the appropriate crystal.

In this circuit, the crystal functions as a parallel resonant LC

Table 5-3. The Parts List for the Crystal Oscillator Circuit of Fig. 5-23.

Vcc	9 volts
Q1	2N2222 or 2N3904 or almost any NPN transistor
R1	22K (22,000 ohms) resistor
R2	56K (56,000 ohms) resistor
R3	27 ohms resistor
R4	1K (1000 ohms) resistor
C1, C3	0.01 μF capacitor
C2, C4	27 pF capacitor
L1	2.5 mH coil
XTAL	crystal—see text

249

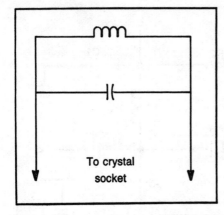

Fig. 5-24. A parallel LC tank can be used in place of the crystal in Fig. 5-23.

circuit. You can prove this to yourself by plugging a parallel resonant LC network (see Fig. 5-24) into the crystal socket.

A fine-tuning control can be added to this circuit by placing a small variable capacitor in series with the crystal, as shown in Fig. 5-25.

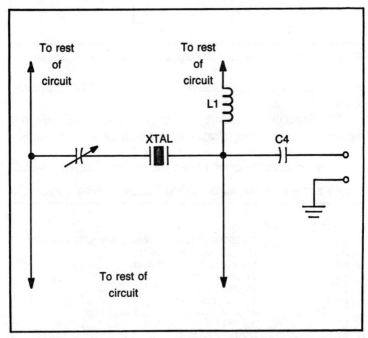

Fig. 5-25. Adding these components will make the output of the crystal oscillator in Fig. 5-23 variable.

EXPERIMENT #6—DESIGN A
MULTIPLE FREQUENCY COLPITTS OSCILLATOR

In working on the design of the Colpitts oscillator earlier in this chapter, we saw that fine-tuning can be accomplished by a small variable capacitor in parallel with the main capacitors in the LC tank. However, unless the variable capacitor is large with respect to the main capacitors, it can only change the resonant frequency by a relatively small amount. In this experiment, we will be designing an oscillator circuit with several overlapping ranges to cover a wider range of output frequencies.

The modified LC tank circuit is shown in Fig. 5-26. S1 is a double-pole, six-throw (DP6T) rotary switch that selects between six C1/C2 combinations (a through f). These capacitors are selected for each of six frequency ranges. The other components in the circuit are selected more or less for the midpoint of the total operating range. The complete circuit is shown in Fig. 5-27.

Let's say VCC has a value of 9 volts, and we want a range of approximately 25 kHz (25,000 Hz) to 60 kHz (60,000 Hz). The required power output is about 10 mW (0.01 watt).

We will use a HEP50 transistor, as in the earlier example. The resistor values are not directly frequency dependent. For our experiment we will use the values calculated in the earlier design example:

$$R1 \quad = \quad 8.2K \ (8200 \ ohms)$$
$$R2 \quad = \quad 470 \ ohms$$
$$R3 \quad = \quad 100 \ ohms$$

Tuning capacitor C3 will be a 365 pF variable capacitor. This value is readily available, because it is often used in portable AM radios. We will assume our tuning capacitor has a capacitance of about 10 pF at its minimum setting.

We will make our preliminary calculations in the middle of the desired band, or about 42 kHz (42,000 Hz).

The load resistance at this frequency works out to roughly:

$$R_L = \frac{VCC}{2 \ P_o} = \frac{9^2}{2 \times 0.01} = \frac{81}{0.02} = 4050 \ ohms$$

Before getting into the parallel resonant tank, let's find the

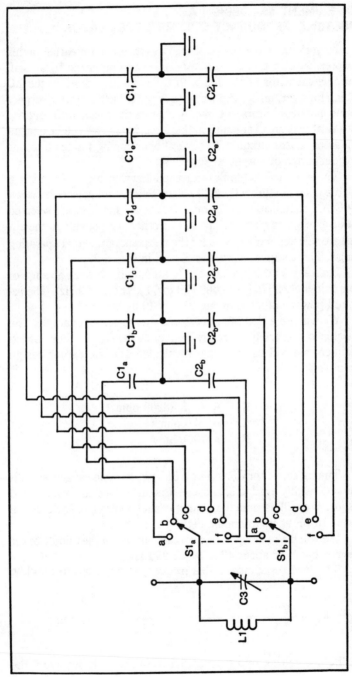

Fig. 5-26. An oscillator can cover a wide range of output frequencies by using a range selector switch.

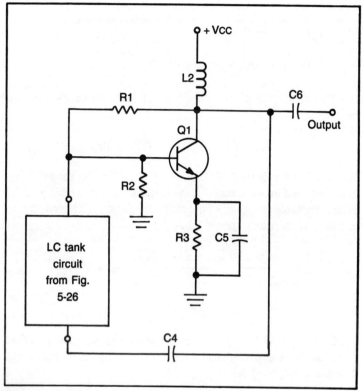

Fig. 5-27. The circuit used for Experiment #6.

values for the remaining components in the circuit, namely, C4, C5, and L2. None of these component values, as you should recall, are particularly critical.

Capacitors C4 and C5 are given equal values. The capacitive reactance for each should be about a tenth of the calculated load resistance, or:

$$X_c = \frac{R_L}{10} = \frac{4050}{10} = 405 \text{ ohms}$$

So, the desired capacitance for these components should be approximately:

$$C = \frac{1}{2 \pi F X_c} = \frac{1}{2 \times 3.14 \times 42000 \times 405}$$

$$= \frac{1}{106876980} = 0.00000000936 \text{ farad}$$

$$= 0.00936 \ \mu\text{F} = 0.01 \ \mu\text{F}$$

The reactance of L2 should be at least 50 times the load resistance:

$$X_L = 50 \times 4050 = 202500 \text{ ohms}$$

We will boost this to 250000 ohms to allow for the range of frequencies used in this circuit. This should keep the reactance of L2 in an acceptable range for all operating frequencies. The inductance of this coil therefore should be about:

$$L = \frac{X_L}{2 \ \pi \ F} = \frac{250000}{2 \times 3.14 \times 42000} = \frac{250000}{263894}$$

$$= 0.95 \text{ henry} = 950 \text{ mH}$$

Now, let's get to the frequency determining components. Coil L1 should have a reactance of 4050 ohms at 42000 Hz. This means its inductance must be:

$$L = \frac{X_L}{2 \ \pi \ F} = \frac{4050}{2 \times 3.14 \times 42000} = \frac{4050}{263894}$$

$$= 0.01535 \text{ henry} \cong 15 \text{ mH}$$

The midrange capacitors should also have a reactance of 4050 ohms at 42000 Hz. This means the total series capacitance of $C1_d$ and $C2_d$ should be:

$$C_T = \frac{1}{2 \ \pi \ FX_c} = \frac{1}{2 \times 3.14 \times 42000 \times 4050}$$

$$= \frac{1}{1068769800} = 0.000000000936 \text{ farad}$$

$$= 0.000936 \ \mu\text{F} = 936 \text{ pF}$$

To determine the individual capacitor values, we need to calculate the transistor's input impedance. First, the collector current is:

$$I_c = \frac{VCC}{2R_L} = \frac{9}{2 \times 4050} = \frac{9}{8100}$$

$$= 0.0011 \text{ amp} = 1.1 \text{ mA}$$

The input impedance can be calculated as follows:

$$Z_i = \beta \times \frac{26}{I_c} = 85 \times \frac{26}{1.1} = 85 \times 23.64$$

$$= 2009 \text{ ohms}$$

The ideal value of capacitor $C2_d$ would therefore be:

$$C2_d = C_T \sqrt{\frac{R_L}{Z_i}} = 936 \times \sqrt{\frac{4050}{2009}}$$

$$= 936 \times \sqrt{2.016} = 936 \times 1.42$$

$$= 1329 \text{ pF}$$

The nearest standard capacitor value is 1500 pF, so that is what we will use for capacitor $C2_d$. This makes the nominal value for $C1_d$:

$$C1_d = \frac{C_T}{1 - C_T/C1} = \frac{936}{1 - 936/15000}$$

$$= \frac{936}{1 - 0.624} = \frac{936}{0.376} = 2489 \text{ pF}$$

If we use a 2500 pF capacitor for $C1_d$, the actual series combination works out to:

$$C_T = \frac{C1_d \times C2_d}{C1_d + C2_d} = \frac{2500 \times 1500}{2500 + 1500}$$

$$= \frac{3750000}{4000} = 937.5 \text{ pF}$$

This is very close to our nominal value of 936 pF, so we're doing just fine.

When we add the effect of the tuning capacitor (C3), the total tank capacitance will range from 947.5 pF (minimum setting) to 1302.5 pF (maximum setting). Let's see what frequency range this gives us. The tank inductance, of course, is a constant 15 mH (0.015 henry).

At the minimum capacitance setting, the resonant frequency is:

$$F = \frac{1}{2\pi\sqrt{LC}}$$

$$= \frac{1}{2 \times 3.14 \times \sqrt{0.015 \times 0.0000000009475}}$$

$$= \frac{1}{6.28 \times \sqrt{0.0000000000142125}}$$

$$= \frac{1}{6.28 \times 0.0000038} = \frac{1}{0.0000237}$$

$$= 42,217 \text{ Hz}$$

At the maximum capacitance setting, the resonant frequency is:

$$F = \frac{1}{2 \times 3.14 \times \sqrt{0.015 \times 0.0000000013025}}$$

$$= \frac{1}{6.28 \times \sqrt{0.0000000000195375}}$$

$$= \frac{1}{6.28 \times 0.0000044} = \frac{1}{0.0000278}$$

$$= 36{,}007 \text{ Hz}$$

With these component values, our oscillator can generate signals with frequencies ranging from 36,007 Hz to 42,217 Hz. We want a full range of 25,000 Hz to 60,000 Hz, so we need to add additional ranges to our circuit, using the method illustrated in Fig. 5-26.

Let's work on our next lower range, which should go up to at least 36,007 Hz. Since we want our ranges to overlap somewhat, we will design this range for a nominal resonant frequency of 37,000 Hz.

We are still using the same coil as before (15 mH), so, since the desired resonant frequency has changed, the load resistance (R_L—reactance at resonance) is also changed. Its new value becomes:

$$X_L = 2 \, \pi \, FL = 2 \times 3.14 \times 37000 \times 0.015$$

$$= 3487 \text{ ohms}$$

Changing the load resistance affects the output power of the circuit, because of this relationship:

$$R_L = \frac{VCC^2}{2 \, P_o}$$

Our nominal output power is 10 mW (0.01 watt). We can rearrange this equation to see how the output power is affected by the changed load resistance:

$$P_o = \frac{VCC^2}{2 \, R_L} = \frac{9^2}{2 \times 3487} = \frac{81}{6974}$$

$$= 0.012 \text{ watt} = 12 \text{ mW}$$

The output power is increased somewhat over its nominal amount, but it is still in the right approximate range. This kind of thing only becomes critical when a fairly high output power is being used. If the maximum power capabilities of the transistor are exceeded, the device could be damaged.

Now, we need to determine the capacitance that will be resonant with the 0.015 henry coil at 300 Hz:

$$C_T = \frac{1}{2\,FX_c} = \frac{1}{2 \times 3.14 \times 37000 \times 3487}$$

$$= \frac{1}{810650290} = 0.000000001234 \text{ farad}$$

$$= 1234 \text{ pF}$$

Solving for the collector current we get:

$$I_c = \frac{VCC}{2R_L} = \frac{9}{2 \times 3487} = \frac{9}{6974}$$

$$= 0.0013 \text{ amp} = 1.3 \text{ mA}$$

Notice that again we have a slight increase over the previous value. The next step is to find the input impedance:

$$Z_i = 85 \times \frac{26}{1.3} = 85 \times 20 = 1700 \text{ ohms}$$

This allows us to solve for the ideal value of capacitor $C2_c$:

$$C2_c = C_T \sqrt{\frac{R_L}{Z_i}} = 1234 \times \sqrt{\frac{3487}{1700}}$$

$$= 1234 \times 2.051 = 1234 \times 1.4322$$

$$= 1767 \text{ pF}$$

If we use a 1800 pF capacitor for $C2_c$, then $C1_c$ must have a value of about:

$$C1_c = \frac{1234}{1 - 1234/1800} = \frac{1}{1 - 0.6856}$$

$$= \frac{1234}{0.3144} = 3924 \text{ pF}$$

This is very close to the standard capacitor value of 3900 pF.

The actual total series capacitance, allowing for the rounded off values is:

$$C_{Tc} = \frac{3900 \times 1800}{3900 + 1800} = \frac{7020000}{5700} = 1232 \text{ pF}$$

At the minimum setting of C3, the total effective capacitance in the tank circuit is equal to 1242 pF, so the output frequency is:

$$F = \frac{1}{2\pi \times \sqrt{0.015 \times 0.000000001242}}$$

$$= \frac{1}{6.28 \times \sqrt{0.00000000001863}}$$

$$= \frac{1}{6.28 \times 0.0000043} = \frac{1}{0.0000271}$$

$$= 36{,}873 \text{ Hz}$$

Notice that this frequency overlaps the lower portion of the previous range.

When C3 is at its maximum setting, the total tank capacitance is 1607 pF. This makes the output frequency equal to:

$$F = \frac{1}{2\pi \times \sqrt{0.015 \times 0.000000001607}}$$

$$= \frac{1}{6.28 \times \sqrt{0.000000000024105}}$$

$$= \frac{1}{6.28 \times 0.0000049} = \frac{1}{0.0000308}$$

$$= 32416 \text{ Hz}$$

Our second range covers output frequencies from 32,416 Hz to 36,873 Hz.

Table 5-4 summarizes ranges b and a. You might want to try the calculations on your own before referring to this table. It is unlikely that you will come up with exactly the same figures, because the resonant frequency points are somewhat arbitrary. Try

Table 5-4. Values for Ranges a and b in Experiment #6.

	RANGE b
NOMINAL FREQUENCY	33,000 Hz
R_L	3110 ohms
Z_i	1527 ohms
C_{T_b}	1528 pF
$C1_b$	5000 pg
$C2_b$	2200 pF
FREQUENCY RANGE	29,867 Hz to 33,136 Hz

	RANGE a
NOMINAL FREQUENCY	30,500 Hz
R_L	2875 ohms
Z_i	1412 ohms
C_{T_a}	1828 pF
$C1_a$	6800 pF
$C2_a$	2500 pF
FREQUENCY RANGE	27,750 to 30,311 Hz

to make sure that you have at least 100 Hz overlap between adjacent ranges.

Notice that the lowest range (a) doesn't quite go down to the desired minimum frequency of 25,000 Hz. Depending on the application, we may want to just live with the narrower range, or design an additional range, although the circuit won't work very well with too wide a range.

It's a good idea to double-check the power dissipation and the collector current for the extreme ranges to make sure that the transistor's specifications will not be exceeded:

$$P_o = \frac{VCC^2}{2R_L} = \frac{9^2}{2 \times 2875} = \frac{81}{5750}$$

$$= 0.014 \text{ watt} = 14 \text{ mW}$$

$$I_c = \frac{VCC}{2R_L} = \frac{9}{2 \times 2875} = \frac{9}{5750}$$

$$= 0.0016 \text{ amp} = 1.6 \text{ mA}$$

That is well within the acceptable range for the HEP50 transistor we are using in this design.

Now we need to move in the other direction and design the two upper ranges (e and f) for frequencies above the midpoint (42,000 Hz).

This works the same as with the lower ranges, of course, except we are now selecting a higher nominal resonant frequency for each range.

Since range d gives output frequencies ranging from 36,007 Hz to 42,217 Hz, let's assign a nominal frequency of 49,000 Hz to range e.

The load resistance (R_L) for this range becomes:

$$R_L = X_L = 2\,\pi\,FL = 2\times 3.14\times 47000\times 0.015$$

$$= 4618 \text{ ohms}$$

This gives us a nominal total capacitance of:

$$C_T = \frac{1}{2\times 3.14\times 47000\times 4618}$$

$$= \frac{1}{1421771700} = 0.000000000703 \text{ farad}$$

$$= 703 \text{ pF}$$

The collector current at 47000 Hz works out to:

$$I_c = \frac{9}{2\times 4618} = \frac{9}{9236} = 0.00010 \text{ amp}$$

$$= 1.0 \text{ mA}$$

The input impedance is:

$$Z_i = 85\times \frac{26}{1.0} = 85\times 26 = 2210 \text{ ohms}$$

Now we can find the individual capacitor values:

$$C2_e = 765\times \sqrt{\frac{4618}{2210}} = 765\times \sqrt{2.09}$$

$$= 765 \times 1.4455 = 1106 \text{ pF}$$

We can round this off and give capacitor $C2_e$ a value of 1200 pF (0.0012 μF). Therefore, capacitor $C1_e$ should have a value of approximately:

$$C1_e \ \frac{76}{1 - 765/1200} = \ \frac{765}{1 - 0.6375} \ \ \frac{765}{0.3625} = 2110 \text{ pF}$$

The nearest standard capacitance value is 2200 pF (0.0022 μF).

Rounding off the capacitance values gives us a combined series capacitance of:

$$C_{Te} = \frac{2200 \times 1000}{2200 + 1000} = \frac{2200000}{3200} = 687.5 \text{ pF}$$

When we add the effects of the parallel tuning capacitor (C3), the total tank capacitance ranges from 697.5 pF to 1052.5 pF.

At the minimum setting of C3, the circuit's output frequency is about:

$$F = \frac{1}{2 \times 3.14 \times \sqrt{0.015 \times 0.0000000006975}}$$

$$= \frac{1}{6.28 \times \sqrt{0.0000000000105}}$$

$$= \frac{1}{6.28 \times 0.0000032} = \frac{1}{0.0000203}$$

$$= 49,204 \text{ Hz}$$

When C3 is readjusted for its maximum capacitance, the output frequency becomes:

$$F = \frac{1}{2 \times 3.14 \times \sqrt{0.015 \times 0.000000000010525}}$$

$$= \frac{1}{6.28 \times \sqrt{0.0000000000157875}}$$

$$= \frac{1}{6.28 \times 0.000004} = \frac{1}{0.000025}$$

$$= 40,056 \text{ Hz.}$$

Table 5-5 summarizes the calculations for the highest range (f). You may come up with somewhat different values.

Our original goal was to design an oscillator with a total range of 25 kHz (25,000 Hz) to 60 kHz (60,000 Hz), and an average output power of approximately 10 mW (0.01 watt).

Our design (a complete parts list is given in Table 5-6) has a frequency range from 27,750 Hz to 59,437 Hz, just a little shy of our original target range.

The output power ranges from a minimum of:

$$P_o = \frac{VCC^2}{2R_L} = \frac{9^2}{2 \times 5655} = \frac{81}{11310}$$

$$= 0.0072 \text{ watt} = 7.2 \text{ mW}$$

This is the power output at the maximum frequency. The maximum output power (on the minimum frequency range) is:

$$P_o = \frac{9^2}{2 \times 2875} = \frac{81}{5750} = 0.014 \text{ watt}$$

$$= 14 \text{ mW}$$

This should be an acceptable design for most applications. If you like, you might want to try to improve the design. The overall frequency range can be increased without adding additional positions to the range selector switch.

Table 5-5. Values for Range f in Experiment #6.

NOMINAL FREQUENCY	60,000 Hz.
R_L	5655 ohms
Z_i	2777 ohms
C_{Tf}	469 pF
$C1_f$	1500 pF
$C2_f$	680 pF
FREQUENCY RANGE	45,025 Hz to 59,437 Hz

**Table 5-6. A suggested Parts List for the
Multiple Range Colpitts Oscillator of Experiment #6.**

Q1	HEP50 NPN transistor
R1	8.2 K resistor (8200 ohms)
R2	470 ohm resistor
R3	100 ohm resistor
L1	15 mH coil
L2	950 mH choke
$C1_a$	6800 pF capacitor
$C1_b$	5000 pF capacitor
$C1_c$	3900 pF capacitor
$C1_d$	2500 pF capacitor
$C1_e$	2200 pF capacitor
$C1_f$	1500 pF capacitor
$C2_a$	2500 pF capacitor
$C2_b$	2200 pF capacitor
$C2_c$	1800 pF capacitor
$C2_d$	1500 pF capacitor
$C2_e$	1200 pF capacitor
$C2_f$	680 pF capacitor
C3	365 pF variable capacitor
C4, C5	0.01 μF capacitors
S1	DP6T rotary switch (see text)

breadboarding system

oscilloscope

frequency counter (optional)

In breadboarding this circuit, you don't really have to bother
with the range selector switch. Simply remove and replace the tank
capacitors for each range. Change components and/or wiring only
with the power disconnected.

Experiment with additional capacitor values. Try changing the
C1:C2 ratio, and see how that affects the output signal.

Monitor the output signal with a frequency counter, or
(preferably) an oscilloscope.

Chapter 6

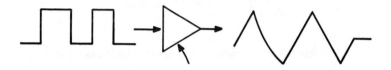

Waveform Generator Circuits

In the last chapter we discussed a number of circuits that generated sine waves. A *sine wave*, you should recall, is a very simple and pure waveform. A typical sine wave is shown in Fig. 6-1.

But sine waves are far from the only usual waveform in electronic circuits. In this chapter we will explore a number of circuits that generate other types of signals.

HARMONICS

The unique feature of sine waves is that they consist of just a single frequency component. The resonant frequency is all there is to the sine wave.

Other waveforms are more complex. They include a number of additional frequency components called *overtones*, or *harmonics*.

A harmonic is an exact integer multiple of the main frequency component or fundamental. For example, let's say we have a signal that contains the first ten harmonics. The main frequency is the fundamental. This is the lowest frequency component in the signal. The second harmonic is two times the fundamental, the third harmonic is three times the fundamental, and so forth. If the nominal frequency of the signal is 100 Hz, the total signal will include the following frequency components:

100 Hz   (fundamental)
200 Hz   (second harmonic)

300 Hz	(third harmonic)
400 Hz	(fourth harmonic)
500 Hz	(fifth harmonic)
600 Hz	(sixth harmonic)
700 Hz	(seventh harmonic)
800 Hz	(eighth harmonic)
900 Hz	(ninth harmonic)
1000 Hz	(tenth harmonic)

(The first harmonic, of course, is the same as the fundamental—1 × F = F.)

This complex signal could be created by combining ten sine waves of the proper frequencies, and the correct amplitude relationships. The relative level of the harmonics also has an effect on the waveshape.

If we change the nominal (fundamental) frequency, the harmonic frequencies will also change, but they will maintain the same relative relationships. For example, if the fundamental frequency in our previous example is changed to 324 Hz, the frequency components in the total signal will be:

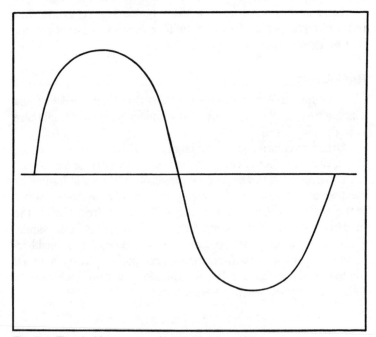

Fig. 6-1. The simplest ac waveform is the sine wave.

324 Hz	(fundamental)
648 Hz	(second—2 × 324)
972 Hz	(third—3 × 324)
1296 Hz	(fourth—4 × 324)
1620 Hz	(fifth—5 × 324)
1944 Hz	(sixth—6 × 324)
2268 Hz	(seventh—7 × 324)
2592 Hz	(eighth—8 × 324)
2916 Hz	(ninth—9 × 324)
3240 Hz	(tenth—10 × 324)

Not all complex signals include all the harmonics. Square waves, for example, contain only the odd harmonics. A typical square wave with a fundamental of 220 Hz will include the following frequency components (up to the tenth harmonic):

220 Hz	(fundamental)
660 Hz	(third)
1100 Hz	(fifth)
1540 Hz	(seventh)
1980 Hz	(ninth)

Notice that all the even harmonics (second, fourth, sixth, eighth and tenth) are omitted. Square waves will be discussed in more depth later in this chapter.

Theoretically, any complex signal can be created by combining enough sine waves at the appropriate frequencies and relative amplitudes. This is called *additive synthesis*. In practical terms, this is usually not very feasible. Besides the obvious expense involved with multiple oscillator circuits, it is extremely difficult, if not completely impossible to keep all the oscillators in precise tune with each other. Even if all the component frequencies could be precisely set, practical oscillator circuits will tend to drift, and there is no way to guarantee they will all drift together. In addition, changing the output frequency would require numerous steps, each of which would require fairly high precision. Moreover, the relative phase relationships between the frequency components can be of significance in many applications.

Fortunately, better methods are generally available. In this chapter we will look at several circuits for generating selected complex waveforms directly.

SOME COMMON WAVEFORMS

Before getting to the actual circuitry, let's take a moment to become familiar with the more commonly used waveforms.

We have already mentioned the square wave, so we will start with it here. A typical square wave is illustrated in Fig. 6-2.

Where the sine wave flows smoothly through a number of different instantaneous amplitudes throughout its cycle, the square wave switches back and forth between just two amplitudes. For half of each cycle it is at a fixed positive amplitude. For the remainder of each cycle, it is at a fixed negative amplitude. These two amplitudes are symmetrical around zero.

An ideal square wave switches instantly between the two levels. A practical circuit, however, would require some finite time to switch between positive and negative output conditions. This causes a distortion of the waveshape called *slew*. A square wave with exaggerated slew is shown in Fig. 6-3. The *switching time* is often referred to as the slew rate.

A square wave is made up of the fundamental and relatively strong odd harmonics. The even harmonics are all omitted.

The relative amplitude of the harmonics follows a simple pattern:

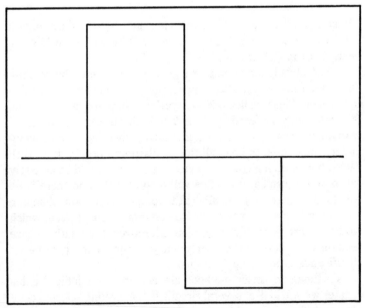

Fig. 6-2. A square wave switches back and forth between two specific levels.

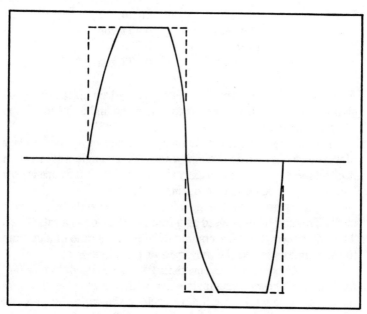

Fig. 6-3. A practical circuit takes some finite time to switch between levels.

$$\text{Amplitude of harmonic X} = \frac{1}{X} \text{ of fundamental}$$

That is, for up to the fifteenth harmonic:

fundamental	1
third harmonic	1/3
fifth harmonic	1/5
seventh harmonic	1/7
ninth harmonic	1/9
eleventh harmonic	1/11
thirteenth harmonic	1/13
fifteenth harmonic	1/15

Let's say, for example, that the amplitude of the fundamental is 5 volts ac RMS. This would give the following amplitudes for each of the harmonics:

third	1.667 volts
fifth	1.000 volt
seventh	0.714 volt

ninth	0.556 volt
eleventh	0.454 volt
thirteenth	0.385 volt
fifteenth	0.333 volt

As you can see, the harmonics very quickly drop off to a fairly insignificant level. This allows us to ignore the higher harmonics in most applications.

The square wave is actually a special case of a broader class of waveforms known as *rectangle waves*. All rectangle waves switch back and forth between two extreme levels. The difference is how long the upper level is held during each cycle.

In the square wave the signal is high for exactly half of each cycle. This can be expressed as a fraction (1/2), or as a ratio (1:2). The ration format is more commonly used. This ratio of high time to total time is called the *duty cycle* of the waveform.

Certainly other duty cycles than 1:2 may be used. Figure 6-4, for instance, shows a rectangle wave with a duty cycle of 1:3. A rectangle wave with a 1:4.5 duty cycle is shown in Fig. 6-5.

The main significance of the duty cycle is the effect it has on the harmonic content of the signal.

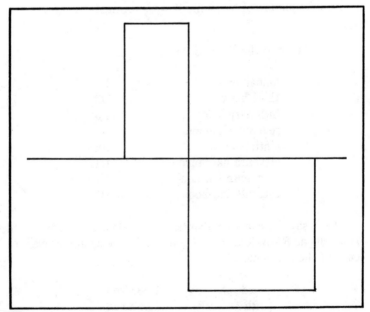

Fig. 6-4. Rectangle wave with a duty cycle of 1:3.

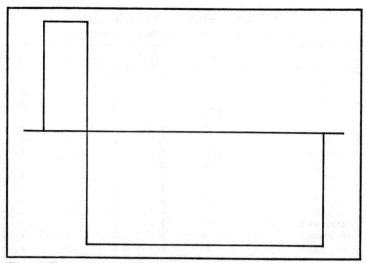

Fig. 6-5. Rectangle wave with a duty cycle of 1:4.5.

The square wave has a duty cycle of 1:2. All harmonics which are multiples of 2 are omitted. This relationship holds true for all duty cycles.

A rectangle wave with a 1:3 duty cycle has every third harmonic absent. The relative amplitudes of the harmonics follow the same relationships described for the square wave:

fundamental	1
second harmonic	1/2
fourth harmonic	1/4
fifth harmonic	1/5
seventh harmonic	1/7
eighth harmonic	1/8
tenth harmonic	1/10
eleventh harmonic	1/11
thirteenth harmonic	1/13
fourteenth harmonic	1/14

The harmonic content of rectangle waves of a number of different duty cycles are compared in Table 6-1.

Another special form of the rectangle wave is the *pulse wave*, illustrated in Fig. 6-6. This is a rectangle wave with a very short high time, so the duty cycle ratio is quite large. The waveform shown has a duty cycle of 1:20. All harmonics are present except

**Table 6-1. The Duty Cycle of a Rectangle
Wave Determines Which Harmonics Are Omitted.**

	1:2	1:3	1:4	1:5	1:6
fundamental	X	X	X	X	X
second	—	X	X	X	X
third	X	—	X	X	X
fourth	—	X	—	X	X
fifth	X	X	X	—	X
sixth	—	—	X	X	—
seventh	X	X	X	X	X
eighth	—	X	—	X	X
ninth	X	—	X	X	X
tenth	—	X	X	—	X
eleventh	X	X	X	X	X
twelfth	—	—	—	X	—
thirteenth	X	X	X	X	X
fourteenth	—	X	X	X	X
fifteenth	X	—	X	—	X

x = harmonic present
– = harmonic absent

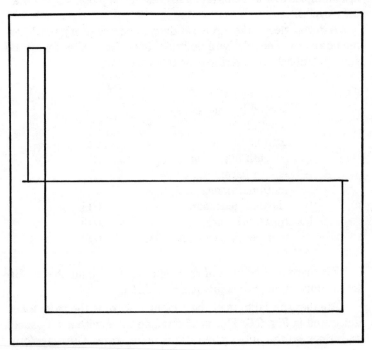

Fig. 6-6. A pulse wave.

for multiples of twenty. For all intents and purposes, all harmonics may be considered to be present. The harmonics that are absent would be so weak if present that they are scarcely missed. Pulse waves are frequently used in applications that require harmonically rich signals.

Another basic waveform is the *triangle wave*, which is shown in Fig. 6-7. This waveform contains all the odd harmonics, and none of the even harmonics, just like the square wave. The harmonics in a triangle wave are much weaker than in a square wave, however. In a triangle wave, the harmonics follow this pattern in their relative amplitudes:

$$\text{Amplitude of harmonic X} = \frac{1}{X^2} \text{ of fundamental.}$$

For the first fifteen harmonics, the relative amplitudes are as follows:

fundamental	1
third	1/9
fifth	1/25
seventh	1/49
ninth	1/81
eleventh	1/121
thirteenth	1/169
fifteenth	1/225

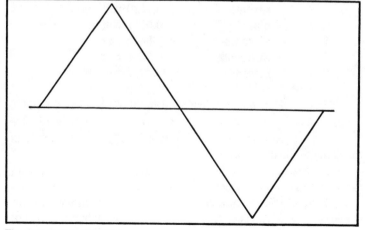

Fig. 6-7. A triangle wave.

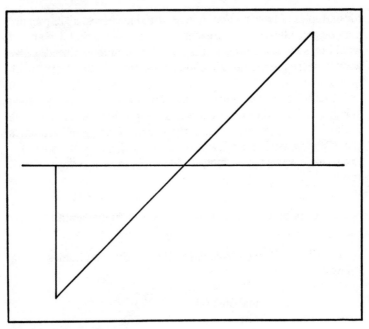

Fig. 6-8. A sawtooth wave.

If the amplitude of the fundamental is 5 volts, the amplitude of the harmonics will be:

third	0.5556 volt
fifth	0.2000 volt
seventh	0.1020 volt
ninth	0.0617 volt
eleventh	0.0413 volt
thirteenth	0.0296 volt
fifteenth	0.0222 volt

Clearly, the harmonics reach an insignificant level very rapidly.

In a sense, a triangle wave is sort of a cross between a sine wave and a square wave, and it is occasionally used for many of the same applications as either.

The final major waveform we will consider here is the *sawtooth wave*, or ramp wave, which is shown in Fig. 6-8.

A sawtooth wave has a very strong harmonic content. All harmonics are included. The relative amplitude of each harmonic drops off exponentially as the harmonic number increases.

RECTANGLE WAVE GENERATORS

Many transistor circuits for generating various types of rectangles, square, and pulse waves have been developed. In this section we will look at just a few.

A super-simple square wave generator circuit is shown in Fig. 6-9. Almost any general purpose NPN transistors may be used in this circuit, but both should be the same type number.

No two transistors are exactly identical. When power is applied to this circuit, one of the two transistors will start to conduct slightly faster than its partner.

Let's say Q1 conducts more heavily than Q2. Q2 will be cut off because Q1's collector voltage is applied to the base of Q2. But, as Q1 continues to conduct more heavily, the collector voltage will start to drop. At some point, the base of Q2 will be at a voltage that allows Q2 to turn on and start conducting. When conduction begins, the collector voltage of Q2 starts out very high. This voltage is fed to the base of Q1, cutting the first transistor off.

The two transistors will switch back and forth, turning each other on and off for as long as power is available to the circuit.

The speed of the switching (output frequency) is determined by the two collector resistors. If these resistors have equal values,

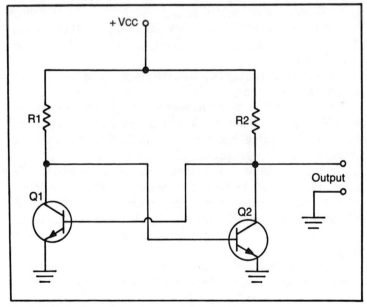

Fig. 6-9. A simple square wave generator circuit.

275

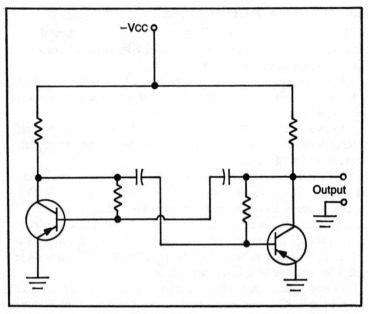

Fig. 6-10. A more sophisticated version of the square wave generator of Fig. 6-9.

the output will be a square wave. Other duty cycle rectangle waves can be achieved with unequal resistances.

The output is shown at the collector of Q2. Another output, which is 180° out of phase with this one can also be taken off the collector of Q1.

The values of the resistors should be kept within the 100 ohm to 2.2K (2200 ohm) range.

This is about the simplest rectangle wave generator circuit you are likely to find. It is not terribly precise in either the output frequency, or waveshape, but it would be perfectly functional for a number of noncritical applications.

A somewhat more sophisticated variation of this circuit is illustrated in Fig. 6-10. In this circuit the output frequency is determined by the charging rates of the capacitors, and the circuit resistances. Again, equal component values will give a square wave.

Notice that PNP transistors and a negative supply voltage (-V_{cc}) are used in this circuit.

The circuit shown in Fig. 6-11 is still more sophisticated. This circuit can be used over a wide range of output frequencies. Fairly clean rectangle waves from 0.5 Hz to 60 kHz (60,000 Hz) can be produced by this circuit.

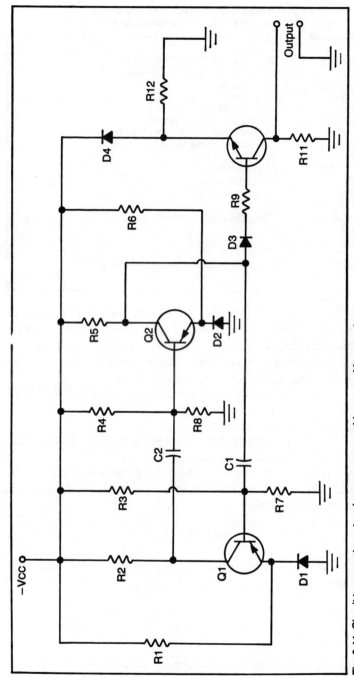

Fig. 6-11. Circuit to generate rectangle waves over a wide range of frequencies.

277

Q1, Q2	2N1303 or similar PNP transistor
Q3	2N1302 or similar NPN transistor
D1, D2	1N463 diode
D3	1N95 diode
D4	1N645 diode
R1, R6	12K (12,000 ohms) resistor
R2	360 ohms resistor — 1 Watt, or better
R3, R4	3.6K (3,600 ohms) resistor
R5	910 ohm resistor
R7, R8	620 ohm resistor
R9	510 ohm resistor
R10	100 ohm resistor
R11	*
R12	5.6K (5,600 ohms) resistor
C1	*
C2	*
V_{cc}	− 12 volts (negative with respect to ground)

When power is applied to this circuit, transistor Q1 starts to conduct. The collection voltage of this transistor rises, charging capacitor C2 through resistor R4. This cuts off Q2, causing its collector to become more negative. This puts a negative charge across C1, speeding up the"on"time of transistor Q3.

Eventually, the charge across C2 will be equal to 63% of V_{cc}. At this point, this capacitor will start to discharge through Q1, which exhibits a low impedance.

Now the second half of the cycle begins, with Q1 and Q2 trading off, and C1 being charged through R3.

The full 0.5 Hz to 60 kHz range can be achieved by using the parts values listed in Table 6-2. Only three component values are left to be calculated.

$$R11 \quad C1 \quad C2$$

The value of resistor R11 will depend on the load being used in the particular application. The formula is:

$$R11 = \frac{V_o}{I_{c3max}I_L}$$

where V_o is the output voltage (which should be no more than 11 volts, if V_{cc} = 12 volts), I_{c3max} is the maximum collector current of transistor Q3, and I_L is the load current.

The output frequency is determined by the values of the two capacitors. The formula for C1 is:

$$C1 = \frac{I_{c1} \times T_L}{0.63V_{cc}}$$

where I_{c1} is the collector current of Q1, and T_L is the time in seconds that the output signal is in its low state per cycle. V_{cc}, of course, is the supply voltage.

The equation for C2 is:

$$C2 = \frac{I_{c2} \times T_H}{0.63V_{cc}}$$

where I_{c2} is the collector current of Q2, and T_H is the time in seconds that the output signal is in its high state per cycle.

The total length of each cycle is the sum of the low time and the high time:

$$T_t = T_L + T_H$$

The frequency is simply the reciprocal of the cycle length:

$$F = \frac{1}{T_t} = \frac{1}{T_L + T_H}$$

Rearranging and combining these equations, we find:

$$T_L = \frac{0.63V_{cc}C1}{I_{c1}}$$

$$T_H = \frac{0.63V_{cc}C2}{I_{c2}}$$

$$T_t = \frac{0.63V_{cc}C1}{I_{c1}} \quad \frac{0.63V_{cc}C2}{I_{c2}}$$

$$F = \cfrac{1}{\cfrac{0.63V_{cc}C1}{I_{c1}} + \cfrac{0.63V_{cc}C2}{I_{c2}}}$$

We can simplify these equations somewhat, by plugging in standard values. VCC is defined in our parts list as 12 volts. Typical values for the two collector currents would be approximately 0.03 ampere (30 mA). Plugging these values into the equations given above, we get:

$$T_L = \frac{0.63 \times 12 \times C1}{0.03} = \frac{7.56C1}{0.03} = 252C1$$

$$T_H = \frac{0.63 \times 12 \quad C2}{0.03} = 252C2$$

$$T_t = 252C1 + 252C2 = 252(C1 + C2)$$

$$F = \frac{1}{252(C1 + C2)}$$

For a square wave, things get even simpler. By definition a square wave signal has equal low and high times:

$$T_L = T_H$$

Therefore:

$$252C1 = 252C2$$

$$C1 = C2 = C$$

$$T_t = 252(C + C) = 252 \times 2 \times C = 504C$$

$$F = \frac{1}{504C}$$

Let's say we want a square wave with a frequency of 1000 Hz. The first step would be to rearrange the frequency equation:

$$C = \frac{1}{504F}$$

Then we can simply solve for C, since the same value is used for both C1 and C2:

$$C = \frac{1}{504 \times 1000} = \frac{1}{504000} = 0.000002 \text{ farad}$$

$$= 2 \ \mu F$$

And that's that.

For other duty cycles, the design is only slightly more complicated. Let's say we need to generate a 12 kHz (12,000 Hz) signal with a duty cycle of 1:5.

First, we simply take the reciprocal of the frequency to find the total cycle time:

$$T_t = \frac{1}{F} = \frac{1}{12000} = 0.0000833 \text{ second}$$

We want the high time to be equal to 1/5 of the total cycle time, so it works out to:

$$T_H = \frac{1}{5} \times T_t = \frac{1}{5} \times 0.0000833 = 0.0000167 \text{ second}$$

Now simple subtraction can give us the low time:

$$T_L = T_t - T_H = 0.0000833 - 0.0000167$$

$$= 0.0000666 \text{ second}$$

Since:

$$T_L = 252C1$$

we can also say:

$$C1 = \frac{T_L}{252} = \frac{0.00000666}{252} = 0.00000026 \text{ farad}$$

$$= 0.26 \ \mu F$$

Similarly, for C2:

$$C2 = \frac{T_H}{252} = \frac{0.0000167}{252} = 0.0000000663 \text{ farad}$$

$$= 0.0663 \ \mu F$$

We can round C1 off to 0.27 μF, and C2 to 0.068 μF.

Let's double-check our work, and see how close we came to our desired output frequency of 12,000 Hz:

$$F = \frac{1}{252(C1 + C2)} = \frac{1}{252(0.00000027 \quad 0.000000068)}$$

$$= \frac{1}{252(0.000000338)} = \frac{1}{0.000085176}$$

$$= 11740 \text{ Hz}$$

The error is due to the cumulative effects of rounding off in the calculations. Still, we came very close—we got 97.8% of our target value.

Let's try one more example. This time we want a 4300 Hz rectangle wave with a duty cycle of 1:3. Working the various equations we come up with the following results:

$$T_t = \frac{1}{F} = \frac{1}{4300} = 0.0002326 \text{ second}$$

$$T_H = \frac{1}{3} \times T_t = \frac{1}{3} \times 0.0002326 = 0.0000775 \text{ second}$$

$$T_L = T_t - T_H = 0.0002326 - 0.0000775 = 0.0001551 \text{ second}$$

$$C1 = \frac{T_L}{252} = \frac{0.0001551}{252} = 0.00000062 \text{ farad}$$

$$= 0.62 \ \mu F$$

$$C2 = \frac{T_H}{252} = \frac{0.0000775}{252} = 0.00000031 \text{ farad}$$

$$= 0.31 \ \mu F = 0.33 \ \mu F$$

Double-checking by recalculating the frequency, we get:

$$F = \frac{1}{252(C1 \cdot C2)}$$

$$= \frac{1}{252(0.00000062 \cdot 0.00000033)}$$

$$= \frac{1}{252 \times 0.00000095} = \frac{1}{0.0002394}$$

$$= 4177 \text{ Hz}$$

This is 97.1% of the desired output frequency of 4300 Hz.

As you can see, it is possible to come very close to the desired frequency with a minimum of mathematical calculations.

If greater precision is required, you could replace resistors R3 and R4 with trimpots to fine-tune the output frequency. I'd suggest using a 2.2K (2200 ohm) resistor in series with a 2.5K trimmer (2500 ohms) for each of these resistors. Calculate the values for C1 and C2 in the usual way, construct (or breadboard) the circuit, and adjust the trimpots while observing the output with an oscilloscope. The disadvantage of this method is that the two controls will interact, affecting both the output frequency and the duty cycle.

A rather novel rectangle wave generator circuit is illustrated in Fig. 6-12. This circuit features three related outputs. Typical outputs from this device are shown in Fig. 6-13. The three times are equal when the following conditions are true:

$$R1 = R3 = R5$$
$$R2 = R4 = R6$$
$$C1 = C2 = C3$$

The circuit will work if these equalities are not true, but the outputs will not be symmetrical.

This type of circuit is particularly useful for sequential switching, especially at low frequencies. As many additional stages as desired can be used.

I recommend that you breadboard this circuit and experiment with various component values.

Almost any general purpose NPN transistor (such as 2N3904s)

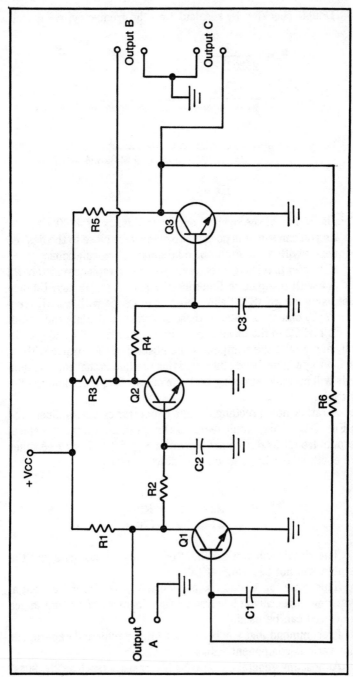

Fig. 6-12. Rectangle wave generator circuit with three outputs.

284

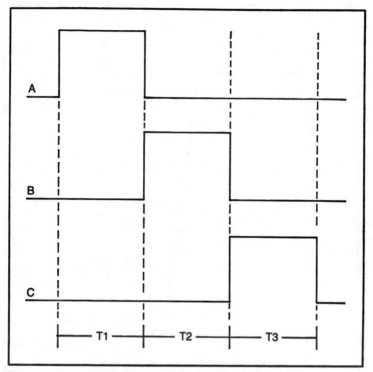

Fig. 6-13. Some typical output signals from the circuit of Fig. 6-12.

can be used. The three transistors should be of the same type for the best results. V_{cc} should be 9 volts.

While this circuit is pretty tolerant of almost any component values, it is probably a good idea to stay within the following ranges in your experimentation:

R1, R3, R5 470 ohms to 3300 ohms
R2, R4, R6 10K (10,000 ohms) to 51K (51,000 ohms)
C1, C2, C3 0.01 μF to 5 μF

Time T1 is determined by R1, R2, and C1. Time T2 is set by R3, R4, and C2. And finally, time T3 is determined by R5, R6, and C3.

SQUARE WAVE CONVERTERS

Sometimes it is desirable to derive a square wave from a sine wave. A number of converter circuits have been developed to ac-

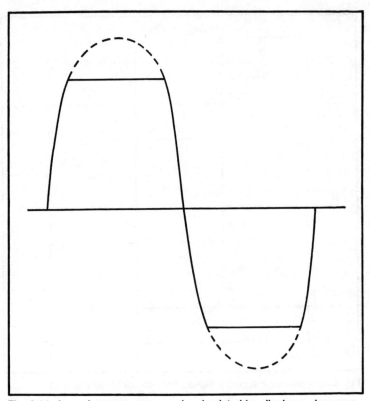

Fig. 6-14. A rough square wave can be simulated by clipping a sine wave.

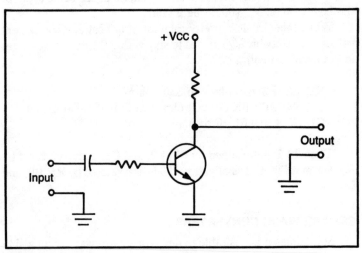

Fig. 6-15. A simple active clipper circuit.

complish this. The output square wave will have the same frequency as the original sine wave.

The simplest approach is to just clip a sine wave, flattening the top and bottom of the waveshape. This results in a fair approximation of a square wave with a poor slew rate, as shown in Fig. 6-14. For more critical applications more precise waveshaping is definitely called for.

A simple active clipper circuit is shown in Fig. 6-15. This is actually a variation on the *Schmitt trigger* (see Chapter 7). When the input signal exceeds a specific trigger voltage, the output goes high. When the input signal drops below a specific trigger voltage, the output goes low.

With a sine wave input, the output will be a square wave that is about 90° out of phase with the original signal. This is shown in Fig. 6-16. Other waveforms at the input will produce other rectangle waves. Almost any input signal will work. Obviously the peak-to-peak voltage of the input signal must exceed the trigger level. If there is a lot of crossing back and forth through the trigger level, the circuit may get confused and the output will become somewhat unreliable.

A somewhat more sophisticated active sine-to-square wave converter circuit is shown in Fig. 6-17. Nothing is terribly critical in this circuit. Potentiometer R4 adjusts the level of the output signal. This may be replaced by a pair of fixed resistors if desired.

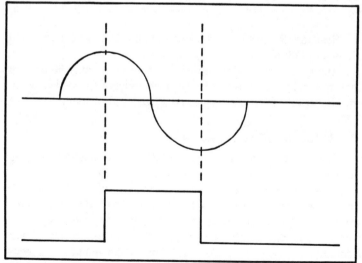

Fig. 6-16. The output of the circuit shown in Fig. 6-15 is shifted 90°.

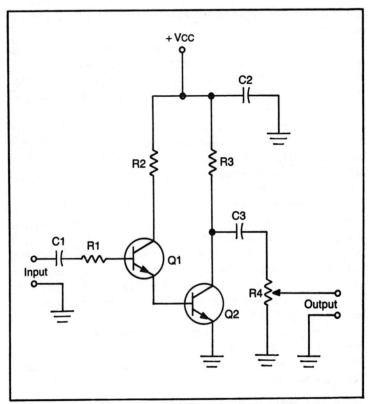

Fig. 6-17. A fairly sophisticated sine wave to square wave converter circuit.

Resistors R2 and R3 should have identical values between 300 ohms and 700 ohms.

Almost any general purpose NPN transistors may be used in this circuit. In fact, if the polarity of the voltage source is reversed, almost any PNP transistors could be used instead.

TRIANGLE WAVE GENERATOR

One way to obtain a triangle wave is to pass a square wave through a low-pass filter, as shown in Fig. 6-18.

A *low-pass filter* does just what it's name implies—it passes low frequencies, but filters out high frequencies. An ideal low-pass filter would completely pass without attenuation any frequency components below its cutoff frequency, and completely block off any frequency components above that point, as shown in the frequency response graph of Fig. 6-19. A practical filter in the real world

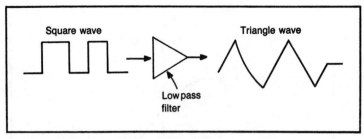

Fig. 6-18. One way to obtain a triangle wave.

will have a more gradual response, like the graph of Fig. 6-20. Frequencies above the cutoff frequency (and just below it) are increasingly attenuated.

The result of this is that higher frequency components are weakened, but not entirely eliminated. A triangle wave, you should recall, has the same harmonic content as a square wave, but the harmonics (which are higher in frequency than the fundamental) are weaker. If a square wave is fed through a low-pass filter with a cutoff frequency just above the fundamental frequency, the output will tend to resemble a triangle wave.

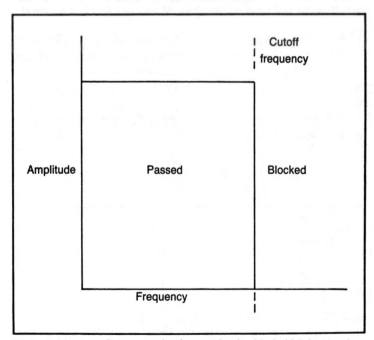

Fig. 6-19. A low-pass filter passes low frequencies, but blocks high frequencies.

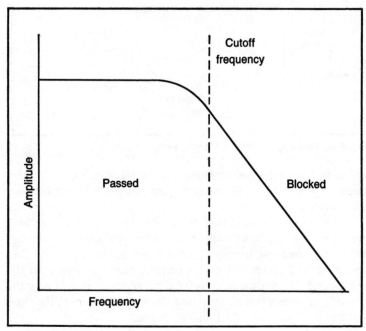

Fig. 6-20. A practical low-pass filter does not have an infinitely sharp cutoff.

A very simple passive low-pass filter is shown in Fig. 6-21. High frequencies see a low reactance through the capacitor, and are shunted to ground. Lower frequencies, however, see a high reactance through the capacitor, so they are fed on through to the output.

The cutoff frequency is determined by the values of the resistor and the capacitor, according to this formula:

$$F = \frac{159000}{RC}$$

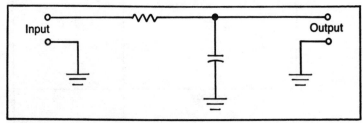

Fig. 6-21. A simple passive low-pass filter can be made from a resistor and a capacitor.

290

For example, let's say we have a low-pass filter made up of a 220K (220,000 ohms) resistor, and a 0.033 μF (0.000000033 farad) capacitor. The cutoff frequency would be equal to:

$$F = \frac{159000}{220000 \times 0.000000033} = \frac{159000}{0.00726}$$

$$= 21,900,826 \text{ Hz} \cong 22 \text{ MHz}$$

As a second example, let's say the resistor is 470K (470,000 ohms), and the capacitor is 0.68 μF (0.00000068 farad). In this case, the cutoff frequency works out to:

$$F = \frac{159000}{470000 \times 0.00000068} = \frac{159000}{0.3196}$$

$$= 497,497 \text{ Hz} \cong 500 \text{ kHz}$$

For best results, the resistance should be chosen to match the input and output impedances as seen by the filter:

$$R = \sqrt{Z_i Z_o}$$

where Z_i is the input impedance, and Z_o is the output impedance.

Once R is selected in this manner, the frequency equation can be rearranged to solve for C:

$$C = \frac{159000}{FR}$$

For instance, let's assume that we need a 100 kHz (100,000 Hz) filter, where the source impedance is 10K (10,000 ohms), and the load impedance is 47K (47,000 ohms). The first step is to find an appropriate value for R:

$$R = \sqrt{10000 \times 47000} = \sqrt{470000000}$$

$$= 21679 \text{ ohms}$$

We can round this off to 22k (22,000 ohms), which is the nearest standard resistance value.

Next, we find the necessary value for C:

$$C = \frac{159000}{100000 \times 22000} = \frac{159000}{2200000000}$$

$$= 0.0000723 \text{ farad} = 72.3 \ \mu\text{F}$$

A 68 μF capacitor can be used. Since the cutoff slope of a passive filter is so gradual, there is a lot of leeway in the component values.

If we really want to know, we can find the actual cutoff frequency for the component values we have selected:

$$F = \frac{159000}{22000 \times 0.000068} = \frac{159000}{1.496}$$

$$= 106,283 \text{ Hz}$$

That's certainly close enough.

Incidentally, we can get just the opposite effect by reversing the position of the two components, as shown in Fig. 6-22. In this case high frequencies pass through the capacitor to the output, but low frequencies are blocked. The frequency response graph for this is shown in Fig. 6-23. This is called, not surprisingly, a *high-pass filter*.

A circuit for generating triangle waves directly is shown in Fig. 6-24. It can be used over a wide range of frequencies simply by using different values for capacitor C1. Typical values for the other components are listed in Table 6-3.

A wide range of values can be used for the capacitor. The larger the capacitance, the lower the output frequency. As Table 6-4 indicates, the circuit can generate frequencies as low as a few Hertz, to several Megahertz (millions of Hertz).

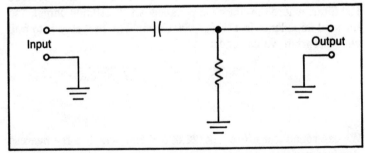

Fig. 6-22. A high-pass filter can be created by reversing the positions of the capacitor and the resistor in a low-pass filter.

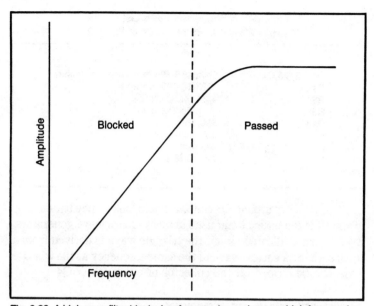

Fig. 6-23. A high-pass filter blocks low frequencies and passes high frequencies.

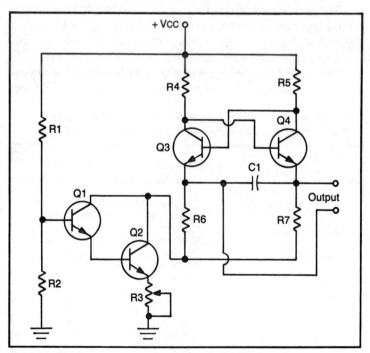

Fig. 6-24. Circuit to generate triangle waves directly.

293

Q1,Q2,Q3,Q4	2N3648 NPN transistor (or similar)
R1	18K (18,000 ohms)
R2	4.2K (4200 ohms)
R3	1K potentiometer
R4,R5	330 ohm
R6,R7	4.7K (4700 ohms)
C1	see text
V_{cc}	12 volts

The circuitry around Q3 and Q4 should look somewhat familiar to you. It is the basic circuit used in most square wave generators. Even in this dedicated circuit the triangle wave is derived from a square wave. A square wave of the same frequency as the triangle wave can be tapped off the collector of either Q3 or Q4.

SAWTOOTH WAVE GENERATORS

Sawtooth waves or ramp waves are used in a wide variety of control applications. For example, a sawtooth wave is used to move the electron beam across the screen in a television set's picture tube.

A basic sawtooth wave generator circuit is shown in Fig. 6-25.

The symbol used for transistor Q1 may be unfamiliar to you. This is a special type of transistor called a UniJunction Transistor,

Table 6-4. The Value of C1 in the Triangle Wave Generator
Circuit of Fig. 6-24 Determines the Output Frequency Range.

C1	MINIMUM	MAXIMUM
100 pF (0.0001 μF)	5.6 MHz (5,600,000 Hz)	10 MHz (10,000,000 Hz)
1000 pF (0.001 μF)	560 kHz (560,000 Hz)	1 MHz (1,000,000 Hz)
0.01 μF	56 kHz (56,000 Hz)	100 kHz (100,000 Hz)
0.1 μF	5.6 khz (5,600 Hz)	10 kHz (10,000 Hz)
1 μF	560 Hz	1 kHz (1,000 Hz)
10 μF	56 Hz	100 Hz
100 μF	5.6 Hz	10 Hz

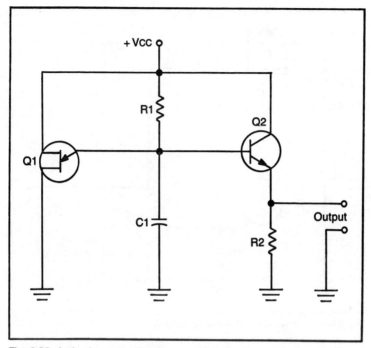

Fig. 6-25. A simple sawtooth wave generator circuit.

or UJT. The UJT and other types of transistors will be discussed in Chapter X.

The circuit shown in Fig. 6-25 is about as simple as it can get— just five components. Capacitor C1 is charged through resistor R1. When the charge on C1 exceeds a specific trigger voltage, the UJT (Q1) is suddenly turned on. When the UJT "fires" it produces a narrow pulse, quickly discharging the capacitor, and the process starts over. Transistor Q2 amplifiers and buffers the changing voltage across the capacitor, giving a sawtooth wave at the output.

Resistor R2 is a load resistor. Its value should be approximately equal to the input impedance of whatever the generator is driving.

The output frequency from this circuit is determined by the R1 = C1 time constant.

A variation of this circuit is shown in Fig. 6-26. This circuit allows a very wide range of output frequencies, depending on the setting of potentiometer R1. For the parts values listed in Table 6-5, the circuit will cover the entire audible range of frequencies (20 Hz to 20 kHz), and extend somewhat into the subaudible and ultra-audible regions.

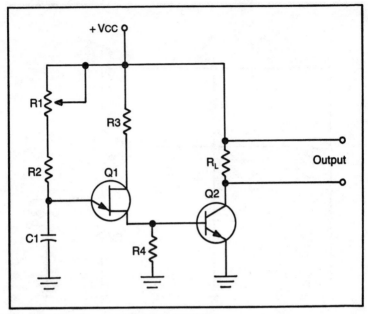

Fig. 6-26. Variable sawtooth wave generator.

This would be a good circuit to experiment with. Breadboard this circuit and experiment with different values for the capacitor. (Remember to disconnect the power before changing any components.) Use values ranging from 0.1 μF to 50 μF and see what

**Table 6-5. Typical Parts Values for the
Sawtooth Wave Generator Circuit of Fig. 6-26.**

Q1	2N2646 UJT
Q2	2N3904 NPN transistor
R1	1 Meg (1,000,000 ohm) potentiometer
R2	3.9K (3900 ohm) resistor
R3	4.7K (4700 ohm) resistor
R4	100 ohm resistor
R_L	selected to match output load
C1	10 μF capacitor
V_{cc}	9 volts

output frequencies you get with a frequency counter or an oscilloscope at the output.

Once again, Q1 is an UJT. Most sawtooth wave generator circuits utilize this type of transistor.

Another variation on the basic sawtooth wave generator circuit is illustrated in Fig. 6-27. The primary difference here is that the capacitor is charged through a constant current source (Q1), rather than through a simple resistor. This allows a straighter, more linear ramp to be generated. A constant current source charges a capacitor linearly over time, rather than the exponential action of charging the capacitor through a resistor.

The charging rate, and thus the output frequency, can be adjusted via potentiometer R2. A smaller resistance here results in a higher output frequency. Using the component values listed in Table 6-6 the output frequency can range from somewhat less than 100 Hz to well over 1000 Hz.

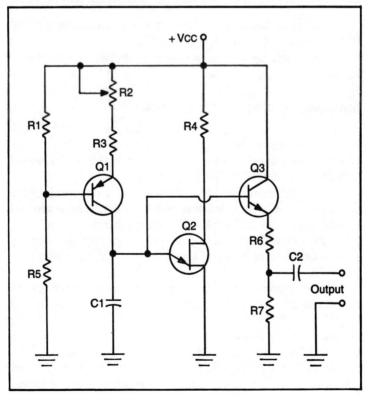

Fig. 6-27. Another sawtooth wave generator circuit.

**Table 6-6. A Suggested Parts List for the
Sawtooth Wave Generator Circuit of Fig. 6-27.**

Q1	2N3906 PNP
Q2	2N2646 UJT
Q3	2N3904 NPN
R1,R4	680 ohm resistor
R2	25K (25,000 ohm) potentiometer
R3,R6	1K (1000 ohm) resistor
R5	3.3K (3300 ohm) resistor
R7	4.7K (4,700 ohm) resistor
C1	0.1 μF capacitor
C2	0.5 μF capacitor
V_{cc}	12 volts

Once again, breadboard this circuit and experiment with different component values—especially C1.

This circuit generates an extremely linear, clean sawtooth wave signal.

UNUSUAL WAVEFORM GENERATORS

Most electronics applications will call for one of the basic waveforms discussed earlier in this chapter. Occasionally, however, more unusual waveshapes may be called for. In this section we will look at a few oddball generator circuits.

The circuit shown in Fig. 6-28 is sort of a double oscillator. Transistor Q2 is a grounded base sine wave oscillator, while both transistors combine to form a square wave generator that turns the sine waves on and off at the output. Consequently, the output is a series of pulses or bursts of a sine wave signal, as illustrated in Fig. 6-29.

When the oscillator is turned on, the time-determining components include C2, R2, D1, and R7. The"off"time is set by R3, R4, R7, and C2. Resistor R4 affects the OFF time independently of the On time. Resistor R2 can be adjusted to control the ON time, but it will also tend to interact with the OFF time, but it will also tend to interact with the OFF time. Therefore, the best approach is to set the ON time first via R2, then adjust R4 for the desired OFF time.

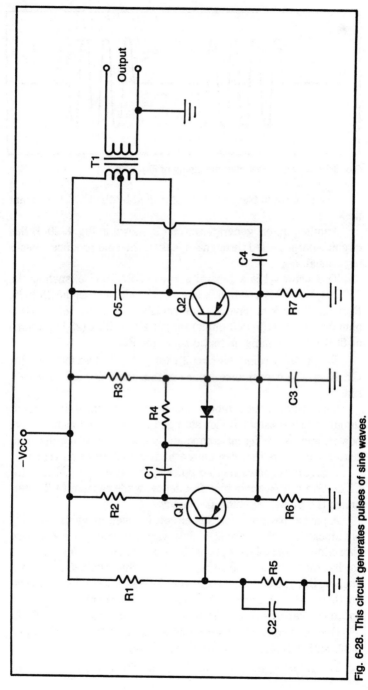

Fig. 6-28. This circuit generates pulses of sine waves.

299

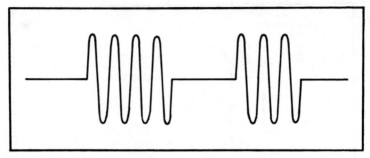

Fig. 6-29. Output signal from the circuit of Fig. 6-28.

The sine wave frequency is controlled by the C5/T1 resonant tank.

Another simple generator circuit is shown in Fig. 6-30. If this circuit's output is fed through a speaker, it would produce a piercing squeal.

Transistor Q1 is a general purpose NPN device, such as the 2N3904, and Q2 is a complimentary PNP unit, such as the 2N3906. Resistor R1 should have a value somewhere in the 68K (68,000 ohms) to 100K (100,000 ohms) range. Use a 25K (25,000 ohms) or 50K (50,000 ohms) potentiometer for R2.

Try breadboarding this circuit with various values for capacitor C1. Keep the capacitance values within the 0.001 μF to 0.1 μF range.

Some component combinations might draw excessive current through the transistors. If they start getting too hot to touch, remove power, and use different component values, or replace the transistors with units that can pass a higher amount of current.

Figure 6-31 shows another pulsed generator circuit. Capacitor C1 determines the main tone frequency, while capacitor C2 determines the pulse switching frequency.

A partial parts list for this circuit is given in Table 6-7. Try breadboarding this circuit using various values for the two capacitors. Typically, C2 should have a much larger value than C1 so that the switching pulse rate is lower than the main tone frequency. But you could get some very interesting results by reversing these ratios. Experiment, and see what you come up with. Try not to use capacitors smaller than 0.05 μF or larger than 500 μF.

One final unusual waveform generator circuit is shown in Fig. 6-32, with a typical parts list in Table 6-8.

The output signal from this circuit is quite unusual. It is recom-

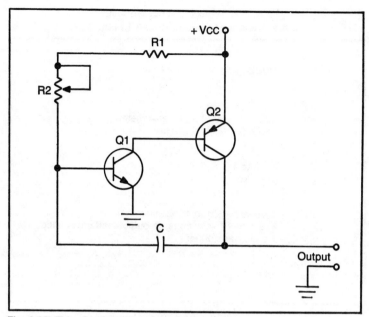

Fig. 6-30. This circuit generates a piercing squeal.

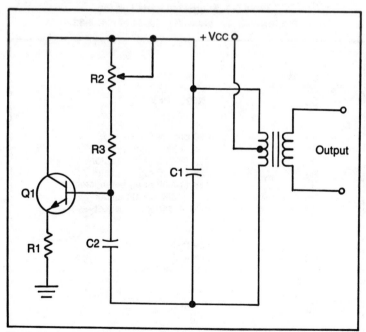

Fig. 6-31. Another pulsed generator circuit.

**Table 6-7. Typical Component Values
for the Unusual Generator Circuit of Fig. 6-31.**

Q1	2N3904 NPN transistor
R1	22 ohm resistor
R2	25K (25,000 ohm) potentiometer
R3	4.7K (4700 ohm) resistor
C1	see text
C2	see text
T1	primary—300 to 1000 ohms secondary—according to output load impedance audio transformer
V_{cc}	9 volt

**Table 6-8. Suggested Parts List for
the Unusual Waveform Generator of Fig. 6-32.**

Q1,Q4	2N2646 UJT
Q2	2N3906 PNP
Q3	2N3904 NPN
R1	680 ohm resistor
R2	18K (18000 ohm) resistor
R3	33K (33000 ohm) resistor
R4	12K (12000 ohm) resistor
R5	10K (10000 ohm) potentiometer
R6,R7,R9	1.8K (1800 ohm) resistor
R8	2.2K (2200 ohm) resistor
C1	see text
C2	see text
C3	0.5 μF capacitor
V_{cc}	12 volt

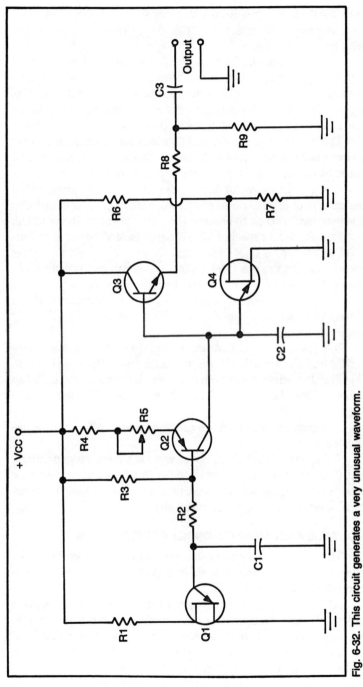

Fig. 6-32. This circuit generates a very unusual waveform.

303

mended that you breadboard this circuit, and experiment with various component values, checking the waveshape at the output with an oscilloscope.

A fairly standard, albeit distorted sawtooth wave is generated by Q1, R1, R2, R3, and C1. If you breadboard this circuit, try placing your oscilloscope probe at the junction of C1 and R2 to see this signal. This sawtooth wave has a rather low frequency due to the size of C1. Try experimenting with other capacitance values here.

The sawtooth wave drives Q2 which functions as a current source charging C2. Unlike the sawtooth wave generator using a constant current source presented earlier in this chapter, this circuit charges the capacitor at an irregular rate, in response to the instantaneous level of the sawtooth wave from Q1. You should also use your oscilloscope to observe the signal at the collector of Q2.

At some point capacitor C2 will have charged up to a sufficient level to trigger UJT Q4 into firing. Q3 and Q4, along with their associated components, form another waveform generation circuit, under the control of the previous two.

The resulting output signal is a very odd waveform. If fed through a speaker, the results probably would not be particularly pleasing.

Breadboard this circuit and experiment with various values for C1 and C2. Usually C1 will be made relatively large (1 μF to 50 μF) so the original sawtooth has a very low frequency. If C2 has a value of about 0.1 to 0.5 μF, the output signal will be in the audio range.

Different resistance values can also have an effect on the output waveform.

We could go on and on about various waveform generators, but that would not be appropriate for this book. My intent here is to demonstrate how basic transistor circuits can be used in a variety of ways to achieve different results.

VOLTAGE-CONTROLLED OSCILLATORS

There is one other type of waveform generator which we should consider before concluding this chapter. This is the voltage-controlled oscillator, or VCO.

All the oscillator and generator circuits had fixed or manually adjustable output frequencies. A VCO allows automatic control of its frequency. The level of an input voltage (or, in some circuits, current) determines the output frequency. Sometimes VCOs are called voltage-to-frequency converters.

A very simple VCO circuit built around a UJT is shown in Fig. 6-33. This circuit is rather unusual in that the control voltage also serves as the circuit's power supply. This circuit can drive a loudspeaker directly.

The output frequency is inversely proportional to the control voltage. That is, the higher the control voltage (providing the maximum ratings of the transistor are not exceeded), the lower the output frequency will be. Typically, the usable range of control voltages will be from about 5 to approximately 30 volts.

Breadboard this circuit and experiment with different values for R1 and C1. R2 should have a value of approximately 100 ohms. This is not too critical, and you can substitute a nearby value if you like. The transistor is a general purpose UJT, such as the 2N2646, or similar device.

The range of output frequencies will be dependent on the values of R1 and C1. Try using resistors with values between 8.2K (8200 ohms) and 62K (62000 ohms). The capacitor can be anywhere from about 0.1 μF to 5μF, or so.

For additional experimentation, try connecting different components between the input and ground. Try a diode, or a capacitor. An electrolytic capacitor with a value around 100 μF produces a particularly striking effect if the output is fed through a loudspeaker.

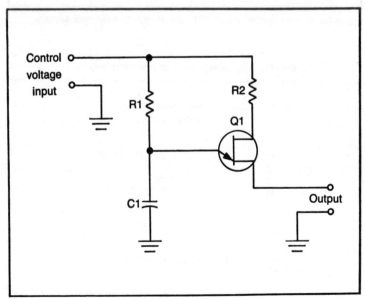

Fig. 6-33. A simple VCO circuit.

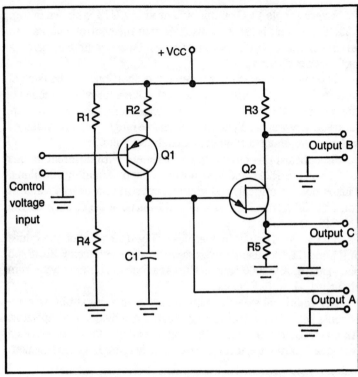

Fig. 6-34. A more sophisticated VCO circuit with three output signals.

**Table 6-9. The VCO Circuit of Fig. 6-34
Can Be Built Using These Component Values.**

Q1	2N1309 PNP
Q2	2N491 UJT
R1	330 K (330,000 ohms)
R2	see text
R3,R5	330 ohms
R4	1 Meg (1,000,000 ohms)
C1	see text
V_{cc}	18 volts to 24 volts

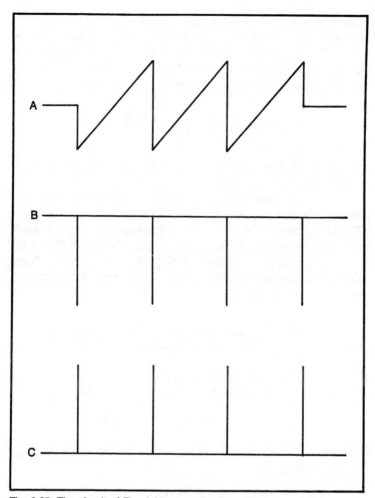

Fig. 6-35. The circuit of Fig. 6-34 generates these three waveforms.

A slightly more advanced VCO circuit is shown in Fig. 6-34. This circuit outputs three different waveforms, a sawtooth wave (output A, from the collector of C1), and two very narrow pulse waves (from Q2—outputs B and C). The three output waveshapes are illustrated in Fig. 6-35.

A partial parts list is given in Table 6-9. These values are not terribly critical, and you may substitute nearby values if you choose.

Resistor R2 and capacitor C1 are the components of most interest to us here, since they determine the range of output frequencies. Try breadboarding this circuit using resistors from 1K (1000 ohms) to 22K (22000 ohms) for R2, and capacitors ranging from about 0.001 μF to approximately 10 μF for C1. The larger the value of these two components, the lower the output frequency will be. The formula for the output frequency is as follows:

$$F = \frac{2}{RC} \left(1 - \frac{V_i}{V_{cc}} \right)$$

where R is the value of R2, C is the value of C1, V_{cc} is the supply voltage, and V_i is the input voltage.

This circuit tends to work best with supply voltages between 18 volts and 24 volts, and input voltages in the 10- to 20-volt range. The input voltage should always be less than the supply voltage. For our calculations, we will assume a supply voltage of 22.5 volts.

Let's try assigning a value of 10K (10000 ohms) to R2 and 0.001 μF to C1. If the control input is 10 volts, the output frequency will be approximately equal to:

$$F = \frac{2}{10000 \times 0.000000001} \left(1 - \frac{10}{22.5} \right)$$

$$= \frac{2}{0.00001} \left(1 - 0.4444 \right)$$

$$= 200,000 \times 0.5556 = 111,111 \text{ Hz}$$

If we raise the control voltage to 15 volts, the output frequency becomes:

$$F = 200,000 \times \left(1 - \frac{15}{22.5} \right)$$

$$= 200,000 \times (1 - 0.6667) = 200,000 \times 0.3333$$

$$= 66,666 \text{ Hz}$$

Increasing the input voltage to 20 volts brings the output frequency down to:

$$F = 200,000 \times \left(1 - \frac{20}{22.5} \right)$$

$$= 200,000 \times (1 - 0.8889) = 200,000 \times 0.1111$$

$$= 22,222 \text{ Hz}$$

As the control voltage increases, the output frequency decreases.

Now, let's leave R2 at 10,000 ohms, but change C1 to 0.5 μF, and find the output frequency for inputs of 10 volts, 15 bolts, and 20 volts:

$$F = \frac{2}{10000 \times 0.0000005} \left(1 - \frac{10}{22.5}\right)$$

$$= \frac{2}{0.005} (1 - 0.4444) = 400 \times 0.5556$$

$$= 222 \text{ Hz}$$

$$F = 400 \left(1 - \frac{15}{22.5}\right) = 400 (1 - 0.6667)$$

$$= 400 \times 0.3333 = 133 \text{ Hz}$$

$$F = 400 \left(1 - \frac{20}{22.5}\right) = 400 (1 - 0.8889)$$

$$= 400 \times 0.1111 = 44 \text{ Hz}$$

If we increase C1 to 10 μF and perform these calculations again, we get output frequencies of:

$$F = \frac{2}{10000 \times 0.00001} \left(1 - \frac{10}{22.5}\right)$$

$$= \frac{2}{0.1} (1 \ 0.4444) = 20 \times 0.5556$$

$$= 11 \text{ Hz}$$

$$F = 20 \times \left(1 - \frac{15}{22.5}\right) = 20 \times (1 - 0.6667)$$

$$= 20 \times 0.3333 = 6.7 \text{ Hz}$$

$$F = 20 \times \left(1 - \frac{20}{22.5}\right) = 20 \times (1 - 0.8889)$$

$$= 20 \times 0.1111 = 2.2 \text{ Hz}$$

Increasing the capacitance decreases the output frequency for a given input voltage. These results are summarized in Table 6-10.

Let's return to the case where C1 = 0.5 μF. With a 10K resistor for R2, we got the following output frequencies for the three inputs:

$$V_i = 10 \qquad F = 222 \text{ Hz}$$
$$V_i = 15 \qquad F = 133 \text{ Hz}$$
$$V_i = 20 \qquad F = 44 \text{ Hz}$$

Now, let's see what happens if we replace resistor R2 with a 3.9K resistor (3900 ohms):

$$F = \frac{2}{3900 \times 0.0000005} \left(1 - \frac{10}{22.5}\right)$$

$$= \frac{2}{0.00195} \times 0.5556 = 1025.64 \times 0.5556$$

$$= 570 \text{ Hz}$$

$$F = 1025.64 \times \left(1 - \frac{15}{22.5}\right) = 1025.64 \times 0.3333$$

$$= 342 \text{ Hz}$$

**Table 6-10. The Value of C1 in Fig. 6-34
Determines the Range of Output Frequencies.**

	C1 = 0.001 μF	C1 = 0.5 μF	C1 = 10 μF
V^i			
10 V	111,111 Hz	222 Hz	11 Hz
15 V	66,666 Hz	133 Hz	6.7 Hz
20 V	22,222 Hz-	44 Hz-	2.2 Hz-

$$F = 1025.64 \times \left(1 - \frac{20}{22.5}\right) = 1025.64 \times 0.1111$$

$$= 114 \text{ Hz}$$

The output frequency is also inversely proportional to the resistance of R2.

Increasing R2, C1, or V_i will reduce the output frequency.

SUMMARY

In this chapter we have briefly looked at just a few waveform generator circuits. We have not gone into as much detail as in the earlier chapters, because, unlike the simpler basic amplifier and oscillator circuits discussed earlier, different applications will call for completely different circuitry, so learning in detail how to design one specific type of sawtooth wave generator, for example, may or may not be useful to the designer, depending on his intended application(s).

However, this chapter should have made you more familiar with waveform generators as a whole, and what to look for in selecting circuits for your specific designs. If you look at the diagrams in this chapter very closely, you will see that they are largely made of simpler subcircuits which are used time and again. Many of the subcircuits should look very familiar to you, since they are variations on circuitry discussed in the first few chapters of this book.

Throughout this chapter it was suggested that you breadboard certain circuits. It seemed more appropriate to discuss these circuits and how to experiment with them within the main body of the text, so there are no separate experiments at the end of this chapter.

Chapter 7

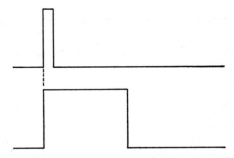

Switching and Pulse Circuits

Most of the circuits described so far in this book have been linear, analog circuits. That is, the output of the transistor varies more or less linearly with the input, and the output is an analog representation of the input signal. This is highly desirable for certain applications, such as amplification and oscillation. In other applications we may not want a continuous range of output levels, but need to switch between clear, unambiguous states. This would be called a *digital circuit*

The simplest digital circuit is basically just a switch, as shown in Fig. 7-1. When the switch is closed, current flows through the resistor, and some positive voltage will be read on the meter. We will call this level E_{high}. When the switch is open, there is no complete current path through the circuit, so no current flows through the resistor, and the meter reads zero. We will call this level E_{low}.

In this circuit the output will always be E_{high}, or E_{low}—never anything in between. The condition of the switch is clearly and unambiguously indicated on the voltmeter. The output is either high or low, on or off, "1" or "0"

We have already dealt with one type of transistor circuit that behaves in this manner—the rectangle wave generators discussed in Chapter 6. In this chapter we will explore in more detail the use of the transistor in switching and pulse circuits.

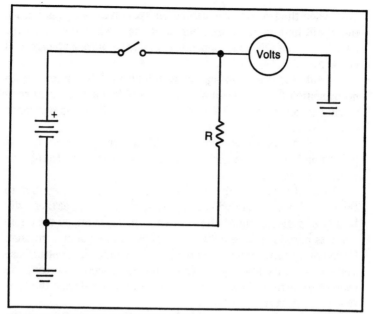

Fig. 7-1. A simple"digital"demonstration circuit.

SWITCHING MODES

To get a transistor to function as an electronic switch, it must be correctly biased. There are three basic biasing schemes, or modes for transistor switching circuits. They are *saturated mode*, the *current mode,* and the *avalanche mode.*

Saturated Mode. In the saturated mode the transistor is turned on by biasing it in saturation. The collector current is limited only by the resistances in the collector and emitter circuits. The voltage across the transistor (known as the *saturation voltage*) is at a minimum. The exact level of the saturation voltage is defined by the collector current and the load resistance.

The OFF condition is achieved by biasing the transistor so that it is cut off—that is, so no collector current flows.

A simple switching circuit is shown in Fig. 7-2. Battery (voltage source) V_{bb} biases the transistor into cutoff when no input signal is present. The base is made negative with respect to the emitter.

If a sufficiently positive voltage is applied to the input, it will overcome this negative bias voltage, switching the transistor on and allowing collector current to flow.

Voltage will be dropped across the load resistor R_L only when there is some collector current. Since there is a collector current only when the transistor is turned on, (positive voltage at input), there will be a voltage drop across R_L only when the transistor switch is in its ON condition. Otherwise, the output voltage will be zero.

Ideally the transistor should switch on and off instantly, with no transition time. This simply isn't possible with practical real-world devices. Some finite time will be required for the transistor to change its output state.

Let's assume we are working with the circuit shown in Fig. 7-2. Initially it is in its OFF condition. There is no collector current, so there is no voltage across R_L.

Now, at some specific time (t_0), a positive pulse is applied to the input. Base current will start to flow right away, but there will be a brief period of time before the emitter/base voltage can climb from its initial negative value through zero, to a positive voltage. Collector current cannot even begin to flow until the emitter/base voltage is at least slightly positive. Then the collector current will require some finite time to reach its maximum level from the starting point of zero.

The turn-on time is considered to be the time from when the positive voltage is first applied to the base (t_0) and the instant

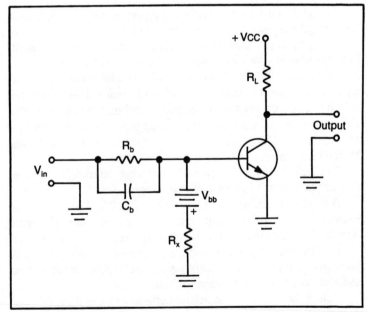

Fig. 7-2. A saturated mode switching circuit.

when the collector current reaches 90% of its maximum value (I1). This is something called *delay time.*

Similarly, the transistor cannot instantly go from saturation to cutoff when the input voltage is removed.

When the positive input voltage is removed, V_{bb} brings the voltage on the base back to a negative level. The base current momentarily goes negative, until the emitter/base voltage goes negative, and the emitter/base junction ceases to conduct. The emitter/base voltage and the collector current remain positive for a brief period after the base is brought down to a negative voltage.

The time it takes for the collector to drop to 90% of its maximum level after the positive input signal is removed is called the *storage time,* and it is determined by the internal capacitances formed within the transistor while it is in saturation. These capacitances are charged when the transistor is turned on, and discharge relatively slowly when the transistor is turned off. Because these internal capacitances are very small, we are ordinarily talking of delays in the microsecond or millisecond range.

Bridging capacitor C_b across resistor R_b increases the switching speed. This capacitor is not absolutely essential for operation of the circuit.

To find the component values in this circuit, you must know the saturation current $I_{c(sat)}$. This value can be easily found via Ohm's Law:

$$I_{c(sat)} = \frac{V_{cc}}{R_L}$$

Resistance R_L, of course, is selected to match the output load being driven by the circuit.

Let's assume the following factors as given:

$$V_{cc} = 9 \text{ volts}$$
$$V_{bb} = 1.5 \text{ volt}$$
$$R_L = 10K \text{ (10,000 ohms)}$$

In this case, the saturation current works out to:

$$I_{c(sat)} = \frac{9}{10000} = 0.0009 \text{ amp} = 0.9 \text{ mA}$$

The value of resistor R_x is found using this form of Ohm's Law:

$$R_x = \frac{V_{bb}}{I_{cbo}}$$

where V_{bb} is the negative base voltage, and I_{cbo} is the collector-to-base leakage current when the transistor is operating at its maximum temperature. This value can be obtained from the specification sheet for the transistor used. If we assume I_{cbo} is 2 μA (0.000002 amp), R_x should have a value of:

$$R_x = \frac{1.5}{0.000002} = 750000 \text{ ohms} = 75K$$

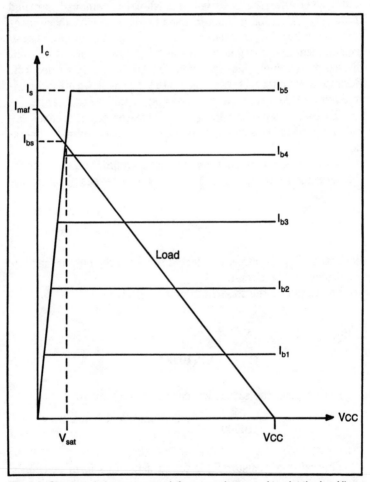

Fig. 7-3. Characteristic curves graph for a transistor used to plot the load line.

Next, we need to look at a characteristic curve graph for the transistor, as shown in Fig. 7-3. Plot the load line for the collector circuit (the solid line marked "LOAD" in the graph). An estimated value for the base current (I_{bs}) is indicated by the point where the load line crosses the transistor's saturation resistance curve. In our example, this is just a little greater than I_{b4}.

We can now find the value of the base resistor (R_b):

$$R_b = \frac{V_{im}}{I_{bs} - I_{cbo}}$$

where V_{im} is the maximum input current.

Let's say I_{bs} works out to 0.225 mA (0.000225 amp), and the maximum input voltage is 2.50 volts. In this case R_b should have a value approximately equal to:

$$R_b = \frac{2.50}{0.000225 - 0.000002}$$

$$= \frac{2.5}{0.000223}$$

$$= 11211 \text{ ohms} \cong 12,000 \text{ ohms} = 12K$$

If C_b is used, its value could be derived mathematically, but it is usually easier just to breadboard the circuit, and experiment with different capacitance values until the best switching speed is obtained.

Current Mode. Higher switching speeds can be obtained if the transistor is not put into saturation when it is turned on. In the current mode the transistor is biased so it operates close to, but not quite in saturation. The collector-emitter voltage is therefore somewhat greater than the saturation voltage of the device.

As in the saturated mode, the transistor switch is turned off, by biasing it so the transistor is in the cut-off (nonconducting) condition.

A typical current mode switching circuit is shown in Fig. 7-4. Notice that it is very similar to the saturated mode circuit of Fig. 7-2.

The battery (V_{bb})/resistor (R_x) combination in the base circuit holds the transistor cut off when there is no input signal. This works essentially the same as the saturated mode circuit described earlier.

The emitter voltage source (V_{ee}) keeps diode D1 turned on (conducting) at all times. Assuming a silicon diode is used, there

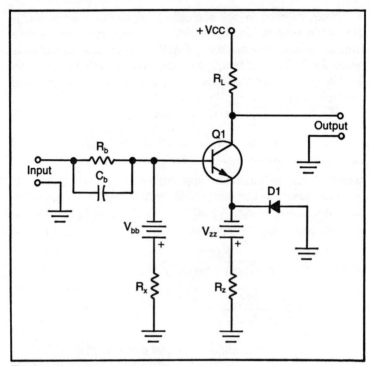

Fig. 7-4. Circuit for a typical current mode switch.

will be approximately 0.7 volt across this component. If the base of the transistor is at ground potential (0 volts between base and ground), only the 0.7 volt across the diode will be between the emitter and the base. This would keep the transistor turned on (conducting).

The base bias voltage (V_{bb}) has to be cancelled out by an input signal with the opposite polarity. When V_{in} is equal to or greater than V_{bb}, the transistor is switched on. When V_{in} is less than V_{bb}, the transistor is cut off.

If V_{ee} was not included in the circuit, we would have practically the same situation as with the saturated mode circuit of Fig. 7-2. When the transistor is turned on by a positive pulse at the input, the collector current will be equal to the beta of the transistor multiplied by the base current. If the base current is large enough, the transistor will be driven into saturation.

The three added components in the emitter circuit (V_{ee}, R_e, and D10 limit the input current, and therefore, the collector current. The maximum current through the transistor is less than its saturation value. The maximum current flow through this circuit's transistor is defined by the values of the components in the emitter circuit.

The diode's polarity prevents any current from the emitter from flowing through it, but the 0.7 volt across the diode does limit the current flow through R_e, which is the maximum current that can flow through either the emitter or the collector in this circuit. The transistor will not be driven into saturation as long as the saturation current value ($I_{c(sat)} = V_{cc}/R_L$) is greater than the current through R_e (V_{ee/R_e}). In practical circuits, the current through the load resistor (R_L) should be limited to a maximum level equal to the current through the emitter resistor (R_e).

In our saturated mode example in the last section, we found that the saturated collector current (the transistor's ON current) was 0.9 mA. For a current mode circuit, we want the emitter limiting current ($I_{el} = V_{ee}/R_e$) to be less than this value. Let's say we want a limit of 0.7 mA (0.0007 amp). If V_{ee} is 1.5 volts, the emitter resistor should have a value of about:

$$R_e = \frac{V_{ee}}{I_{el}} = \frac{1.5}{0.0007} = 2143 \text{ ohms}$$

If we use a standard 2.2K (2200 ohm) resistor, the maximum collector current/emitter current will be approximately:

$$I_{cmax} = \frac{1.5}{2200} = 0.00068 \text{ amp} = 0.68 \text{ mA}$$

The transistor will not be put into saturation when it is turned on.

Because the transistor does not have to work quite so hard (conduct as much current) in its ON state, it can switch between states considerably faster than in a comparable saturated mode switching circuit.

Avalanche Mode. The third switching mode is also the fastest. This is the avalanche mode. The ON and OFF states of the transistor switch are kept within the breakdown portion of the transistor's operating curve. Faster switching can only be obtained with special devices such as hot-carrier, pin, snap-off, or tunnel diodes.

Saturated mode and current mode switching circuits require a specific base voltage to be maintained to hold the transistor off (or on). In an avalanche mode circuit, however, a brief pulse is all that is needed to hold the transistor in either ON or OFF state.

MULTIVIBRATORS

If we combine two switching transistors circuits so that when one is on, the other will be off, we have a type of circuit known as a multivibrator. Obviously a multivibrator can only have one of two possible output conditions at any time (ignoring the brief switching time).

There are three primary types of multivibrators. They are *monostable* multivibrators, *bistable* multivibrators, and *astable* multivibrators.

The Monostable Multivibrator. The monostable multivibrator has one stable output state, which may be either high or low, depending on the specific design used. When an input signal is received, the output will momentarily switch over to the opposite state for a specific period of time, then return automatically to the original stable state.

The output pulse is of a fixed length depending on component values within the monostable multivibrator circuit itself. The length of the input pulse used as a trigger signal is irrelevant. For this reason, monostable multivibrators are occasionally referred to as *pulse stretchers*. This function is illustrated in Fig. 7-5.

Monostable multivibrators can be used to slow down very rapid signal pulses, and for a wide variety of timing applications.

A typical monostable multivibrator circuit is shown in fig. 7-6.

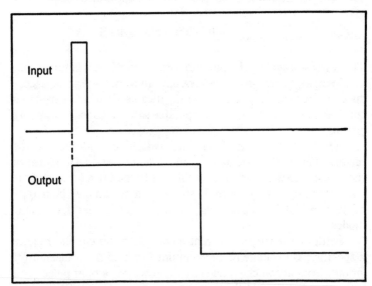

Fig. 7-5. A monostable multivibrator used as a pulse stretcher.

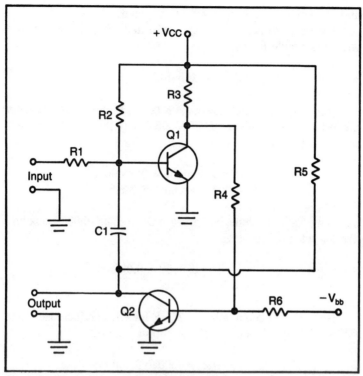

Fig. 7-6. A typical monostable multivibrator circuit.

Ordinarily, when there is no input signal, transistor Q1 is held on because of the biasing current to the base through resistor R2 from the $+V_{cc}$ supply voltage. A negative voltage ($-V_{bb}$) is applied to the base of transistor Q2, through resistor R6, keeping this device in the cutoff state.

This state of affairs remains unchanged as long as there is on signal at the input. If a positive pulse is applied to the input, nothing will happen either. A negative input pulse, however, will overcome the positive bias on the base of Q1, cutting that transistor off. The collector voltage will jump up to $+V_{cc}$. This positive voltage is now fed to the base of transistor Q2 through resistors R3 and R4. $+V_{cc}$ has a higher potential than $-V_{bb}$, so the negative bias on Q2 is overcome, and this transistor is now turned on.

Even if the input pulse is now removed, Q2 will remain on until capacitor C1 has time to discharge through R2. Once the capacitor has discharged sufficiently, Q2 will be again cutoff, and Q1 will be turned on. The circuit returns to its stable state until another input pulse is received at the input.

The time required for the capacitor to discharge can be found with this simple formula:

$$T = 0.69R2C1$$

If resistor R2 has a value of 220K (220,000 ohms), and capacitor C1 has a value of 0.05 μF (0.00000005 farad), the output pulse will have a length of:

$$T = 0.69 \times 220000 \times 0.00000005$$
$$= 0.00759 \text{ second} \cong 8 \text{ mS}$$

If we leave R2 alone, but increase C1 to 25 μF (0.000025 farad), the output time becomes:

$$T = 0.69 \times 220000 \times 0.000025$$
$$= 3.795 \text{ seconds}$$

Now, if we keep a value of 25 μF for C1, and increase R2 to 680K (680,000 ohms), we get an output time of:

$$T = 0.69 \times 680000 \times 0.000025$$
$$= 11.73 \text{ second}$$

Increasing either the resistance or the capacitance will increase the time constant.

The length of the input pulse has absolutely no effect on the length of the output pulse. Once the cycle is begun, its timing will be determined solely by the RC time constant of the circuit.

By selecting the proper resistance/capacitance combination, almost any length of output pulse may be generated. Large component values give long output times, on the order of several seconds. Small component values give short output pulses of just a millisecond (thousandth of a second), or even less.

Output pulses of various lengths can be obtained by using a potentiometer, or other variable resistance for R2. For example, let's say we use a 1 Meg (1,000,000 ohms) potentiometer for R2, and a 0.2 μF (0.0000001 farad) capacitor for C1. Assuming the potentiometer has a resistance of 100 ohms at its minimum setting, the minimum output pulse time would be:

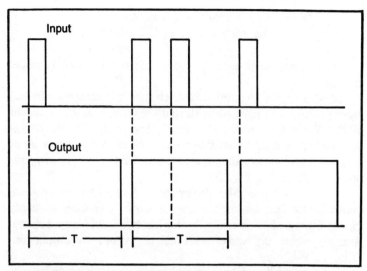

Fig. 7-7. Most monostable multivibrators ignore additional input pulses while the output is active.

$$T = 0.69 \times 100 \times 0.0000001$$
$$= 0.0000069 \text{ second} = 0.0069 \text{ ms} = 6.9 \ \mu s$$

At the maximum setting of the potentiometer, the output pulse would have a length of:

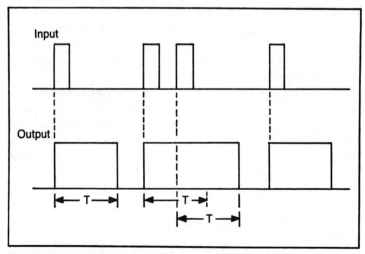

Fig. 7-8. Some monostable multivibrators will start their timing cycle over if a new input pulse is received before the previous output pulse is completed.

$$T = 0.69 \times 1000000 \times 0.0000001$$
$$= 0.069 \text{ second} = 69 \text{ ms} = 69000 \ \mu\text{s}$$

A very large range of output times can be obtained with a single circuit.

Most monostable multivibrator circuits will ignore additional input pulses that might occur during the output pulse, as shown in Fig. 7-7. Other monostable multivibrator circuits are retriggerable. The output pulse time is set back to zero if a new input pulse is received before the circuit has timed out. This type of operation is illustrated in Fig. 7-8.

The Bistable Multivibrator. Since the monostable multivibrator has one stable output state, you have probably guessed that a bistable multivibrator has two stable output states. A single circuit may hold either a high output or a low output condition indefinitely.

Let's say we have a bistable multivibrator circuit which initially has a low output. It will stay low until a trigger pulse is sensed at the input, at which time the output of the bistable multivibrator will go high. This condition will be held constant until a second trigger pulse is received, causing the output to return to the low state. On the next input pulse the output will go high again. This can continue indefinitely. The output switches back and forth between output states on each input pulse.

Because of the back and forth output operation, bistable multivibrators are often called *flip-flops*.

In a sense, a bistable multivibrators is a simple memory device. It"remembers"what state it was last placed in. It will continue to remember until a new trigger pulse changes its state, or the power source is interrupted.

A circuit that remembers if it's supposed to be on or off doesn't seem as though it would be particularly useful, but, in fact, there are a great many important applications for bistable multivibrators. The programmable memories (RAM) in most computers are essentially made of a great many flip-flops arranged in meaningful patterns. This is why the computers use the *binary numbering system*, even though it is very clumsy for us humans. In the binary system, there are just two digits—0 and 1. Since a bistable multivibrator has two possible output conditions, the connection should be pretty obvious.

Typical input and output signals for a bistable multivibrator are shown in Fig. 7-9. If the input is a regular square wave, the output will also be a square wave with exactly half the frequency of the input signal. This is because two input pulses equal one output pulse:

INPUT PULSE#1	BEGIN OUTPUT PULSE#1
INPUT PULSE#2	END OUTPUT PULSE#1
INPUT PULSE#3	BEGIN OUTPUT PULSE#2
INPUT PULSE#4	END OUTPUT PULSE#2
INPUT PULSE#5	BEGIN OUTPUT PULSE#3
INPUT PULSE#6	END OUTPUT PULSE#3
INPUT PULSE#7	BEGIN OUTPUT PULSE#4
INPUT PULSE#8	END OUTPUT PULSE#4
INPUT PULSE#9	BEGIN OUTPUT PULSE#5
INPUT PULSE#10	END OUTPUT PULSE#5

and so on.

If the input is a 500 Hz square wave, the output will be a 250 Hz square wave.

In Fig. 7-9 the output pulses are initiated by the rising portion of the input pulses. In some circuits, the output pulses are triggered by the falling portion of the input pulses. This may be more desirable in certain applications. Usually the choice between rise-triggering and fall-triggering is pretty arbitrary, and simply depends on the design used.

Many bistable multivibrator circuits have two inputs. One input, which is often labeled "SET," forces the output to a high level.

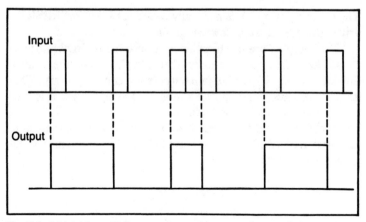

Fig. 7-9. A bistable multivibrator reverses its output state each time an input pulse is received.

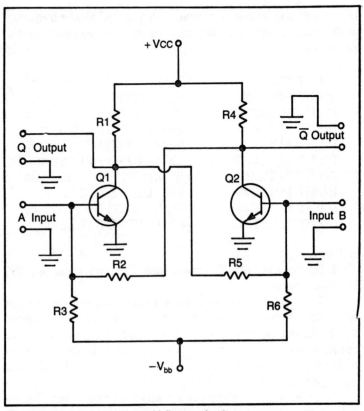

Fig. 7-10. A typical bistable multivibrator circuit.

The other input, which is generally called "CLEAR" or "RESET," drives the output to its low condition.

Two outputs are often included in bistable multivibrator circuits. The main output is usually labelled "Q", and the other is identified as "$\overline{Q}$"(not Q). These two outputs are complimentary. That is, they are always in opposite states. If Q is high, then $\overline{Q}$ is low, and if Q is low, then $\overline{Q}$ is high.

A fairly typical transistor bistable multivibrator circuit is illustrated in Fig. 7-10. The two halves of this circuit are mirror images of each other. That is:

$$Q1 = Q2$$
$$R1 = R4$$
$$R2 = R5$$
$$R3 = R6$$

326

In this circuit one of the transistors will always be cutoff, and the other will be in saturation. The input pulses reverse the states of the two transistors.

To keep one of the transistors in saturation, its base current should be equal to:

$$I_b = \frac{I_{c(sat)}}{\dfrac{V_{cc}}{R1 + R2} - \dfrac{V_{bb}}{R3}}$$

When power is first applied to this circuit, one of the transistors will start to conduct before its partner. That transistor will go into saturation, and the other will be cutoff. For our discussion, we will assume that initially Q1 is saturated, and Q2 is cutoff.

At this point, the collector of Q1 (Q output) is near ground potential (low), and the collector of Q2 ($\overline{Q}$ output) is at V_{cc} (high).

Assuming there is no input signal, the base of Q2 is fed Q1's collector signal through R5, and $-V_{bb}$ through R6. Since the voltage at the collector of Q1 is about 0, the voltage applied to the base of Q2 is negative, holding this device in cutoff.

Similarly, assuming there is no input signal, the base of Q1 is fed Q2's collector signal through R2, and $-V_{bb}$ through R3. Since the potential of Q2's collector is equal to V_{cc}, which is significantly larger than V_{bb}, the overall voltage on the base of Q1 will be a relatively large positive value. This keeps Q1 in saturation. As you can see, this circuit is in a stable condition as long as there is no input signal.

If a positive pulse is now applied to input A, nothing will happen, since this pulse will just be added to the already positive voltage on the base of Q1.

On the other hand, if a positive pulse is applied to input B, it will overcome the negative bias voltage on the base of Q2, and this transistor will be driven into saturation. Its collector voltage drops to zero, so the controlling voltage on the base of Q1 goes negative, cutting this device off.

Now we have just the opposite condition we started with. The collector of Q1 is equal to V_{cc} (output Q is high), and the collector of Q2 has a near zero voltage (output $\overline{Q}$ is low). This new state of affairs will remain constant, even if the pulse at input B is removed.

Now, if another positive pulse is applied to input B, nothing will happen, since Q2 is already in saturation.

If a positive pulse is then applied to input A, Q1 will be turned on and Q2 will be turned off. We will be back to our original starting condition. As long as power is applied, the circuit will "remember" which input was last activated.

The Astable Multivibrator. A multivibrator has two possible output conditions. In a monostable multivibrator one of those output conditions is stable. In a bistable multivibrator, both are stable. What is left for the third type of multivibrator—the astable?

In an astable multivibrator neither of the output conditions is stable. The output keeps switching back and forth between states without a controlling input. The output signal of an astable multivibrator is illustrated in Fig. 7-11. Does this signal look familiar to you? It certainly should. It is a square wave. Yes, an astable multivibrator is simply a square wave generator.

A typical astable multivibrator circuit is shown in Fig. 7-12. Notice how similar it is to the monostable and bistable circuits discussed earlier in this chapter.

Once again, when power is first applied to the circuit, one of the transistors will start to conduct a bit ahead of the other. That transistor will quickly reach saturation, and the other will be cut off.

We will start our discussion with the assumption that Q1 is on, and Q2 is off. The collector voltage of Q2 is equal to V_{cc}, and the collector voltage of Q1 is near zero.

Capacitor C1 is charged to a value just under V_{cc} by the collector voltage of Q2. The polarity of this charge will be as indicated in the figure.

During the previous half-cycle, capacitor C2 was charged with the polarity shown. As C1 is now charged, C2 starts to discharge. When C2 has been discharged, a base current flows through Q2 due to the current flowing through R4 from V_{cc}. This positive base

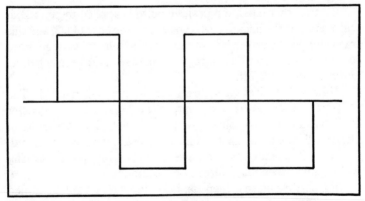

Fig. 7-11. An astable multivibrator is simply a square wave generator.

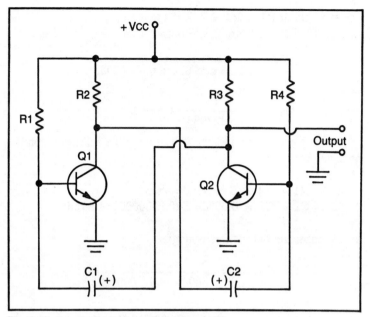

Fig. 7-12. A typical astable multivibrator circuit.

current turns Q2 on, pulling its collector and the positive side of C1 down to ground potential.

Because of the polarity of the charge on C1, the base of Q1 is negative, cutting this transistor off. During the time Q1 is off, its collector is at V_{cc}, charging capacitor C2 with the polarity shown. As C2 is being charged, C1 is being discharged through R1, and the whole process is reversed.

For a square wave output (duty cycle of 1:2), the two halves of the circuit should be symmetrical:

$$R1 = R4$$
$$R2 = R3$$
$$C1 = C2$$

The length of an entire cycle would be equal to:

$$T_t = 1.38R1C1$$

Since frequency is the reciprocal of cycle time:

$$F = \frac{1}{T_t} = \frac{1}{1.38R1C1}$$

For example, if R1 = R4 = 27K (27,000 ohms), and C1 = C2 = 0.33 μF (0.00000033 farad), the output frequency would be:

$$F = \frac{1}{1.38 \times 27000 \times 0.00000033}$$

$$= \frac{1}{0.0122958} \cong 81 \text{ Hz}$$

For a second example, we will assume the following component values:

R1 = R4 = 5.6K (5600 ohms)
C1 = C2 = 0.0022 μF (0.0000000022 farad)

In this case, the output frequency is:

$$F = \frac{1}{1.38 \times 5600 \times 0.0000000022}$$

$$= \frac{1}{0.000017} \cong 58{,}818 \text{ Hz}$$

If unequal component values are used, other duty cycles can be achieved. The equations become only slightly more complicated.

The high level output time (T_h) equals:

$$T_h = 0.69R4C2$$

and the low level output time (T_l) is:

$$T_l = 0.69R1C1$$

The total cycle time, of course, is simply the sum of these two times:

$$T_t = T_h + T_l$$

The output frequency, of course, is the reciprocal of the total cycle time:

$$F = \frac{1}{T_t} = \frac{1}{T_h + T_l}$$

$$= \frac{1}{0.69R4C2 + 0.69R1C1}$$

For instance, if we assume the following component values:

$$R1 = 1K \ (1000 \ ohms)$$
$$R4 = 4.7K \ (4700 \ ohms)$$
$$C1 = 0.1 \ \mu F \ (0.0000001 \ farad)$$
$$C2 = 0.068 \ \mu F \ (0.000000068 \ farad)$$

the high level output time will be:

$$T_h = 0.69 \times 4700 \times 0.000000068$$
$$= 0.00022 \ second = 0.22 \ ms$$

and the low level output time is:

$$T_l = 0.69 \times 1000 \times 0.0000001$$
$$= 0.000069 \ second = 0.069 \ ms$$

The total cycle time is therefore:

$$T_t = 0.22 + 0.069 = 0.289 \ ms$$
$$= 0.000289 \ second$$

We can figure the duty cycle by comparing the high output time (T_h) to the total cycle time (T_t):

$$T_h : T_t$$
$$0.22 : 0.289$$
$$1 : 1.313$$

Finally, the output frequency is:

$$F = \frac{1}{T_t} = \frac{1}{0.000289} = 3454 \ Hz$$

We will consider just one more example before moving on. This time, we will assume the following component values:

$$R1 = 3.9K \ (3900 \ ohms)$$
$$R4 = 15K \ (15000 \ ohms)$$
$$C1 = 0.22 \ \mu F \ (0.00000022 \ farad)$$
$$C2 = 0.22 \ \mu F \ (0.00000022 \ farad)$$

Notice that this time the capacitors have identical values, but the resistors have differing values.

The output frequency resulting from this combination of components is equal to:

$$F = \frac{1}{0.69 \times 15000 \times 0.00000022 + 0.69 \times 3900 \times 0.00000022}$$

$$= \frac{1}{0.002277 + 0.000592} = \frac{1}{0.002869}$$

$$= 348.5 \text{ Hz}$$

The circuit can cover a wide range of output frequencies by the proper selection of component values.

A secondary output puts out a signal that is 180° out of phase with the main output. This can be taken from the collector of Q1. When one output goes high, the other will go low, and vice versa. There are many applications in which such complimentary outputs can come in very handy.

Chapter 8

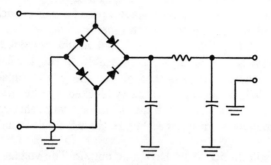

Power Supply Circuits

Every circuit requires some kind of power supply to operate. A few passive circuits, such as the passive filters mentioned in Chapter 6, "steal" their power from the signal they operate on, but the majority of circuits (including all circuits with active components, such as transistors) require a definite voltage source.

In many of the circuit diagrams throughout this book, batteries were indicated as voltage sources. Batteries are convenient sources of clean dc voltages, but they can get rather expensive and inefficient, especially in applications that draw a great deal of current. Generally it is more practical to use the ac power of your standard house wiring. This power source has a lot to offer. It is convenient, fairly cheap, and readily available most of the time. The biggest disadvantage is that it is in an ac power source, and most circuits require a dc power source.

To use the ac electrical power from your outlets, you need a circuit that will convert the ac power into a dc voltage. Such a circuit is called a power supply. Actually a power supply does not really supply power—it simply converts one type of electrical power (ac) into another (dc).

HALF-WAVE RECTIFIERS

The simplest possible power supply circuit is a diode. A diode, as you should recall, is a single PN junction. Electrical current can pass through the junction in one direction, but not the other. An ac signal is continuously reversing direction (polarity). If we pass an ac signal through a diode, as shown in Fig. 8-1, half on each cycle will be deleted. The output polarity will be constant.

Obviously this simple"circuit"doesn't do a very good job of converting ac to dc. The output voltage varies for half of each input signal, and is zero the rest of the time.

Adding a capacitor across the output, as illustrated in Fig. 8-2 will improve the output signal considerably. During the half cycle the input signal can pass through the diode, the capacitor is charged. When the input signal is blocked by the diode, the capacitor slowly discharges. If the capacitor is large enough, it will not be completely discharged before the next output half cycle begins.

We still do not really have a dc output. The voltage sort of "wobbles" around its normal value. A half-wave rectifier can be used to power some noncritical circuits, but many circuits require a more stable voltage source.

Another big disadvantage of the half-wave rectifier is that half of the input power is simply wasted, since half of each input cycle is blocked by the diode.

A practical half-wave rectifier circuit is shown in Fig. 8-3. Resistor R2, along with capacitors C1 and C2, acts as a low-pass

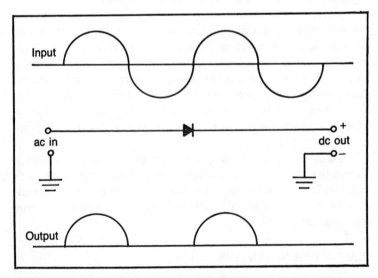

Fig. 8-1. Current can pass through a diode in only one direction. Half-wave rectification.

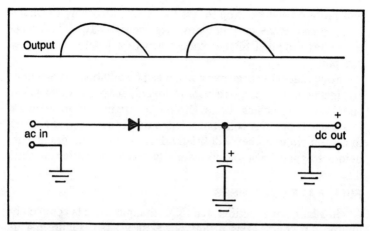

Fig. 8-2. Placing a capacitor across the output smooths the output signal.

filter to smooth out the "dc" voltage a little better than a single capacitor could. There will still be considerable voltage fluctuations (or "ripple"), but they won't be quite as pronounced.

R1 is a surge resistor. It is used to protect the diode from any sudden increase of current drawn through the circuit (perhaps due to a short circuit in the load). The surge resistor typically has a fairly small value, so normally the voltage drop across it is rather small. But an increase in the current drawn through this resistor

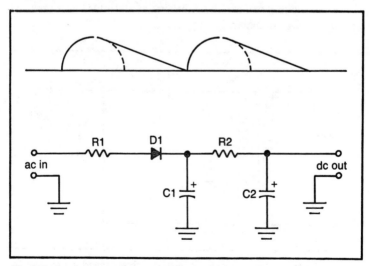

Fig. 8-3. A practical half-wave rectifier circuit.

will cause its voltage drop to rise too, since, according to Ohm's Law, voltage equals current times resistance (E = IR). This brings the voltage applied to the diode down to a level it can safely dissipate.

Sometimes the surge resistor is fused for additional protection.

Incidentally, a surge resistor is not only for protection against circuit defects such as shorts. Often it is also needed for more or less normal operating conditions. Often when power is first applied to certain circuits, they will briefly draw a great deal of current, before dropping back to a considerably smaller operating current.

FULL-WAVE RECTIFIERS

In a half-wave rectifier, half of each input cycle is completely unused. Of course, this means power is wasted. Figure 8-4 illustrates a more efficient type of power supply circuit, called a full-wave rectifier. Notice that a full-wave rectifier requires a center-tapped transformer.

If the center tap of a transformer's secondary winding is grounded, the lower half of the secondary winding will carry an ac signal that is equal to, but 180° out of phase with the signal through the upper half of the secondary winding. As a result of this, when diode D1 is passing a positive half cycle, diode D2 is blocking a negative half cycle. And, when D1 is blocking a negative half cycle, D2 will be passing a positive half cycle. One of the diodes is conducting and one is nonconducting at all times. This gives us the output signal shown in Fig. 8-5.

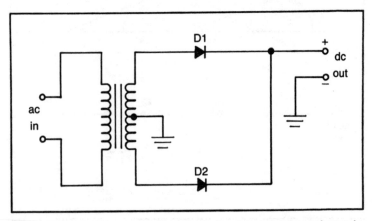

Fig. 8-4. A basic full-wave rectifier circuit. A center-tapped transformer is required.

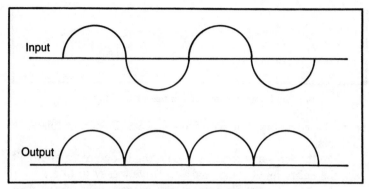

Fig. 8-5. Full-wave rectification uses the full input cycle.

Notice that besides wasting less input power (the entire input signal is put to work), the full-wave rectifier has a more constant output than a half-wave rectifier.

As in the case of the half-wave rectifier circuit, the ripple in the output signal of a full-wave rectifier circuit can be reduced by using filter capacitors, as shown in Fig. 8-6.

Notice that both the positive and negative output lines require their own filter capacitor, and they are isolated from the ac ground.

A full-wave rectifier signal is easier to filter than a half-wave rectifier signal, since the capacitor won't have as much time to discharge before a new rising half cycle recharges it. This gives a smoother, more dc-like output. See Fig. 8-7.

BRIDGE RECTIFIERS

A bridge rectifier circuit, like the one shown in Fig. 8-8 com-

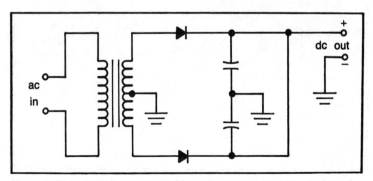

Fig. 8-6. Filter capacitors can provide a smoother output from a full-wave rectifier.

337

Fig. 8-7. The filtered output signal from a full-wave rectifier.

bines the advantages of both half-wave rectifiers and full-wave rectifiers.

Like the full-wave rectifier circuit, the bridge rectifier circuit uses the entire input cycle, and is relatively easy to filter.

Like the half-wave filter circuit, the bridge rectifier circuit does not require a bulky and expensive center-tapped transformer. While a bridge rectifier circuit does require four diodes (instead of the two used in the full-wave rectifier), it is still usually more economical for semiconductor circuits than most center-tapped transformers. The circuit also requires less space, as a rule, and produces less heat.

Ready-made bridge rectifiers (four individual diodes in a bridge arrangement within a single package) are readily available too.

Another similarity to the half-wave rectifier is that one of the bridge rectifier's output lines may be at ground potential, as shown in the schematic.

At any point of the input cycle, two of the diodes in the bridge are conducting, and two are reverse-biased. During the positive half

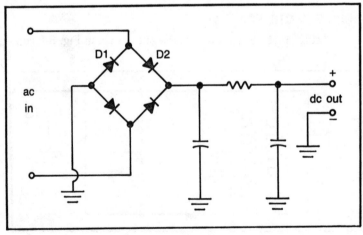

Fig. 8-8. A bridge rectifier uses four diodes. Two are forward-biased, and two are reverse-biased at all times.

cycle, diodes D1 and D4 are effectively open circuits. The equivalent circuit during the positive half cycle is illustrated in Fig. 8-9. Similarly, Fig. 8-10 shows the equivalent circuit during the negative half cycle.

VOLTAGE REGULATION

All the basic power supply circuits discussed so far share a common problem. Their output voltage is not a constant dc value. Some circuits are very critical, and need a very precise, smooth voltage.

A lot of the ripple problem can be reduced by using filters, but some ripple will always get through. Still, ripple is not the only problem. The output voltage will also vary with changes in the ac input voltage, and the load resistance/impedance of the circuit being driven by the power supply.

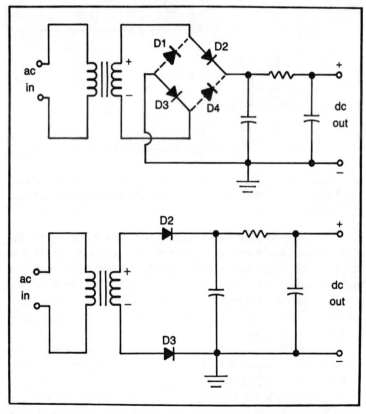

Fig. 8-9. During the positive half-cycle, diodes D1 and D4 are effectively out of the circuit.

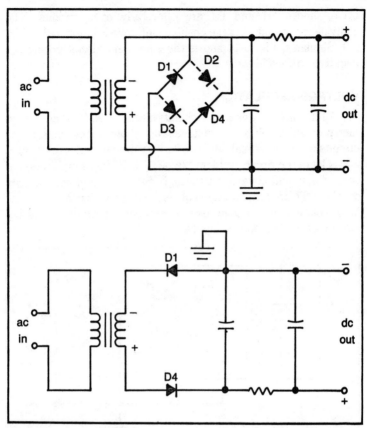

Fig. 8-10. During the negative half-cycle, diodes D2 and D3 are effectively out of the circuit.

You might assume that if the ac input voltage comes from the electric company's wiring that it will be a constant 120 volts ac (RMS). This simply isn't the case at all. The electricity from your outlets tends to vary quite a bit, depending on weather conditions, and the current load being placed on the lines by you and your neighbors. If it stays within ±10% of the nominal value, count yourself very lucky.

I've seen the line voltage drop as low as 80 volts in peak drain times (such as the dog days of summer, when every air conditioner in town is running full tilt).

In addition, the load being driven by the power supply circuit is rarely a simple fixed dc resistance. As the current drawn by the load circuit changes, so does the effective load resistance seen by the power supply circuit. This can affect the output voltage considerably.

The solution is to include some voltage regulation circuitry.

Voltage regulators in IC form are widely used today but the designer should be familar with other techniques that may be better choices under other circumstances.

Zener Diode Regulation. The simplest voltage regulation device is the zener diode.

An ordinary diode conducts when forward-biased, and blocks the current flow when reverse-biased. This action is shown in the graph of Fig. 8-11. Notice the slight deviation from the nominal performance around zero. This is due to the voltage drop across the semiconductor diode. This is approximately 0.7 volt for silicon diodes.

The action of a zener diode is illustrated in the graph of Fig.

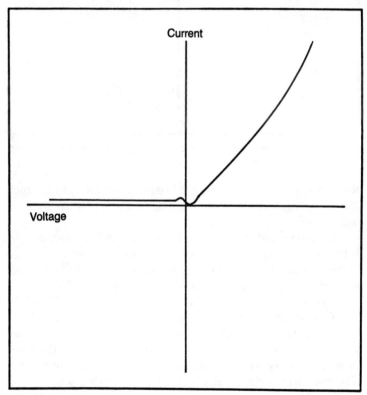

Fig. 8-11. An ordinary diode passes current when it is forward-biased, and blocks current when it is reverse-biased.

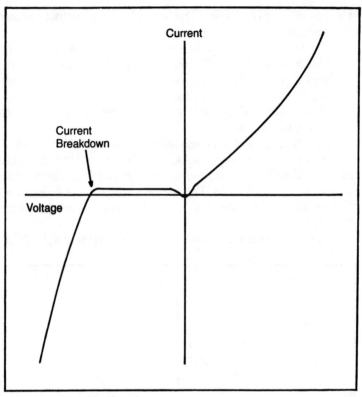

Fig. 8-12. A zener diode starts to conduct heavily if the reverse bias voltage exceeds the breakdown voltage.

8-12. When the diode is forward-biased, it behaves pretty much like an ordinary diode. When the zener diode is reverse-biased is when it shows its difference.

If the reverse-biased voltage is below a specific point, current is blocked, as with ordinary diodes. But when this specific voltage (called the breakdown voltage) is exceeded, even by a very small amount, the zener diode starts to conduct very heavily.

The schematic symbol for a zener diode is shown in Fig. 8-13.

To understand the operation of a zener diode as a voltage regulator, consider the circuit shown in Fig. 8-14. Potentiometer R_x acts a voltage divider limiting the voltage being fed to the zener diode voltage regulator circuit, which consists of an input resistor (R_i), the load resistance (R_L), and the zener diode itself.

The breakdown voltage of a zener diode is fixed for the individual device. A number of standard voltages are available. For

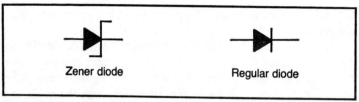

Fig. 8-13. Schematic symbol for a zener diode.

our discussion, we will assume we are working with a 5.6 volt zener diode.

If the potentiometer is set so the input voltage (indicated on voltmeter A) is zero, the output voltage (as read on voltmeter B) will also be zero, of course. If no voltage is applied to the circuit, obviously no voltage can appear.

When the input voltage is raised to some positive level—we'll say one volt—resistor R_i (which is included to limit the current through the zener diode to a safe level) drops a small amount of the source voltage. With one volt of reverse bias, the diode is in its cutoff region. It does not conduct, and the circuit behaves as if the zener diode wasn't there at all. The output voltage (B) will be one volt, less the small amount dropped across R_i.

If we continue raising the input voltage, the output voltage will rise in step with the input. This will continue to the point when the source voltage exceeds the *breakdown voltage* of the zener diode

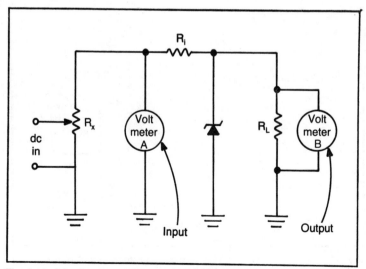

Fig. 8-14. A basic zener diode voltage regulation circuit.

(5.6 volts in our example). This point is also known as the *avalan-che point.*

Once the breakdown voltage is passed, the zener diode will conduct very heavily. In fact, it will try to pass more current than it can handle without damage. This is why the input resistor (R_i) is needed in the circuit. It limits the current through the diode to a safe level.

Since the zener diode conducts any voltage above 5.6 volts to ground, the output voltmeter (B) will read about 5.6 volts, even if the input voltage is increased to 6 volts, 7 volts, 8 volts, or even higher.

The zener diode protects the load circuit from any surges in the power source, which could damage delicate components.

The zener diode will also help regulate the voltage against changes in the load resistance. The voltage remains fairly constant, regardless of the amount of current drawn by the load.

To understand how this works, let's see what happens when we vary the load resistance (R_L) with the zener diode left out of the circuit. The modified experimental circuit is illustrated in Fig. 8-15.

Assuming the input resistance (R_i) is a constant 500 ohms, Table 8-1 shows the current and the voltage drop across the load resistance for various values of R_L. Notice that the voltage can vary quite a bit as the load resistance changes.

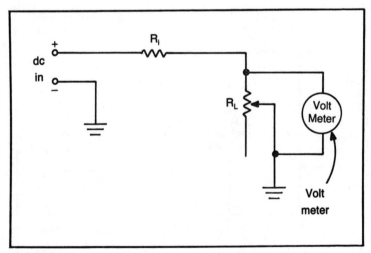

Fig. 8-15. Adding this circuit to the voltage regulator of Fig. 8-14 allows you to experiment with changes in the load resistance.

344

| | Source voltage = 7 volts | | | |
| | Input resistance (R_i) = 500 ohms | | | |
R_L	R(total)	I (mA)	Voltage Drop—R_i	Voltage Drop—R_L
100	600	11.7	5.83	1.17
200	700	10.0	5.00	2.00
300	800	8.75	4.38	2.62
400	900	7.78	3.89	3.11
500	1000	7.00	3.50	3.50
600	1100	6.36	3.18	3.82
700	1200	5.83	2.92	4.08
800	1300	5.38	2.69	4.31
900	1400	5.00	2.50	4.50
1000	1500	4.67	2.33	4.67
1200	1700	4.12	2.06	4.94
1400	1900	3.68	1.84	5.16
1600	2100	3.33	1.67	5.33
1800	2300	3.04	1.52	5.48
2000	2500	2.80	1.40	5.60
2200	2700	2.59	1.30	5.70
2400	2900	2.41	1.21	5.79
2600	3100	2.26	1.13	5.87
2800	3300	2.12	1.06	5.94
3000	3500	2.00	1.00	6.00
3500	4000	1.75	0.87	6.13
4000	4500	1.56	0.78	6.22

With the zener diode in the circuit, as shown in Fig. 8-14, however, the output voltage will be held to a fairly constant 5.6 volts, as long as the input voltage is at least slightly higher than this.

Since the load resistance is in parallel with the zener diode, the load voltage will be equal to the drop across the diode. Thus, this is a simple, but effective voltage regulation circuit.

If the effective load resistance drops for any reason (that is, if the load circuit starts to draw more current), this will cause an increase in the voltage drop across R_i, corresponding to the decreasing voltage drop across R_L. This decreases the voltage supplied to the diode itself. As less voltage is applied to the diode, it will draw less current. This means that the voltage drop across R_i must decrease, due to Ohm's Law. The output voltage is forced to stabilize at the level defined by the zener diode's breakdown point.

A zener diode can do a fair job of voltage regulation in low current applications (although it can't do much if the input voltage drops below the desired value). When larger currents, or more precise regulation are required, more sophisticated voltage regulation circuits are called for.

Series Regulator. One type of voltage regulator circuit using transistors is the series regulator. Dc flows from the unregulated input voltage through a transistor to the load.

A simple series regulator circuit is shown in Fig. 8-16. Current flows through resistor R1 and the zener diode. Thus, a fixed voltage appears across the zener.

Meanwhile, the current through R1 is also flowing through the base/emitter junction of the transistor. A fixed voltage is developed across this junction, due to the semiconductor material used. For ordinary silicon transistors, this junction voltage will be about 0.7 volt. This is sufficient to turn on the transistor. The voltage at the emitter (referenced to ground) is equal to 0.7 volt plus the voltage across the zener diode (its breakdown voltage). This voltage is placed across the load resistance (R_L) and is constant, even if the supply voltage or the load resistance changes.

One of the major advantages of this circuit is that only a relatively small amount of current flows through the zener diode. Resistor R1 limits this current. An inexpensive low-power zener diode can be used, regardless of the current requirements of the load circuit. The current supplied to the load flows through the transistor. Clearly, if this type of circuit is used in fairly high power applications, the transistor should be a heavy-duty, high-power type. Adequate

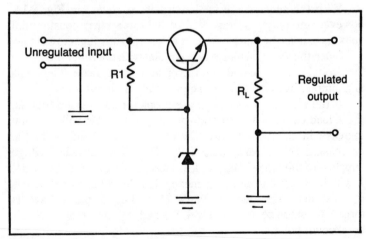

Fig. 8-16. A simple series regulator circuit.

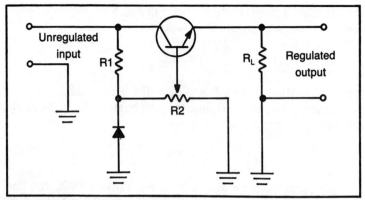

Fig. 8-17. Series regulator circuit.

heat sinking is a definite must in most power supply applications. Always using a hefty heat sink on all power supply transistors is a very good habit to get into.

A simple modification of the circuit of Fig. 8-16 allows the regulated output voltage to be manually variable. The modified circuit is illustrated in Fig. 8-17.

In this case the output voltage (across R_L) is equal to the voltage between the potentiometer's wiper (center connection) and ground plus the base/emitter junction voltage (0.7 volt). The maximum output voltage is just slightly over the breakdown voltage of the zener diode.

If this approach is used, the value of resistor R1 becomes more critical. It must be selected to pass enough current to keep the transistor turned on at all times, but not to ever drive it into saturation.

Figure 8-18 shows an improvement on the basic series regulator circuit. Instead of a single transistor, a Darlington pair is used. This presents a higher impedance to the zener diode.

In the basic circuit of Fig. 8-16, the impedance seen by the zener diode is equal to the load resistance multiplied by the transistor's beta:

$$Z = \beta \ R_L$$

For instance, if the beta of the transistor is 85, and the load resistance is 10K (10,000 ohms), the impedance seen by the zener diode will be:

$$Z = 85 \times 10000 = 850,000 \text{ ohms} = 850K$$

If a Darlington pair is used, as shown in Fig. 8-18, the impedance seen by the zener diode will increase to:

$$Z = \beta1\beta2R_L$$

where $\beta1$ is the beta of Q1, and $\beta2$ is the beta of Q2.

In most cases, the two transistors will be identical, so the equation can be rewritten as:

$$Z = \beta^2R_L$$

For the example given above ($\beta = 85$, $R_L = 10000$), the impedance seen by the zener diode in the circuit with the Darlington pair is:

$$Z = 85^2 \times 10000 = 7225 \times 10000$$

$$= 72,250,000 \text{ ohms} = 72.25 \text{ Megohms}$$

This is certainly a significant increase.

Another improvement is added to the circuit in Fig. 8-19. Better regulation can be achieved if a constant current is applied to

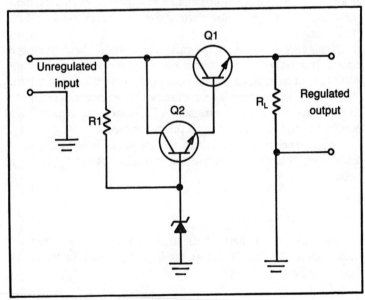

Fig. 8-18. Series regulator circuit using a Darlington pair.

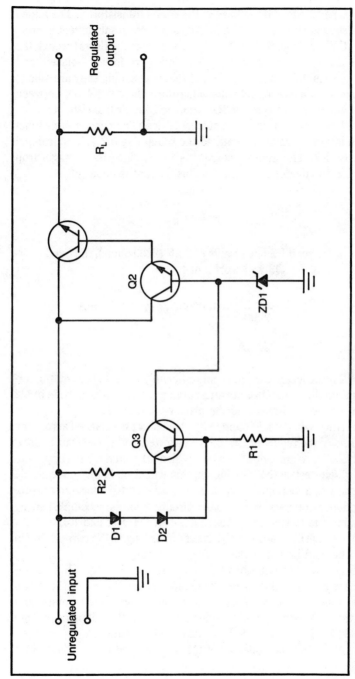

Fig. 8-19. Constant current source added for better voltage regulation.

the zener diode and the base of the series transistor(s). The added components (Q3, R2, D1, and D2) form a constant current source.

Current from the unregulated voltage source flows through the two diodes and resistor R1. The diodes are forward-biased, and so there is 0.7 volt across each of them (assuming they are silicon diodes), for a total of 1.4 volts. Therefore, there is 1.4 volt between the upper end of resistor R2, and the base of transistor Q3.

The transistor is also a silicon device. The base/emitter junction is turned on at 0.7 volt, so the remaining 0.7 volt is dropped across R2. The emitter current is fixed, since the voltage drop across the diodes is constant. This current is equal to:

$$I_e = \frac{0.7}{R2}$$

For instance, if R2 has a value of 2.2K (2200 ohms), the fixed emitter current for Q3 will work out to:

$$I_e = \frac{0.7}{2200} = 0.00032 \text{ amp} = 0.32 \text{ mA}$$

$$= 320 \ \mu A$$

The collector current is approximately equal to the emitter current, so the amount of current applied to ZD1 and the base of Q2 can be selected by using the proper value for R2.

Now, what would happen if a dead short was placed across the output of the regulator circuit? The shorted load would start to draw an excessive amount of current, which Q1 would try to supply until it self-destructed. Obviously, this would be highly undesirable for any application. In Fig. 8-20 we have added short circuit protection to our series regulator circuit. This is accomplished by adding the following components—Q4, D3, R3, and R4.

Ordinarily, transistor Q4 is cut off, as long as the current drawn by the load is within the a acceptable range.

If the current through the load resistance (R_L) is increased, so will the current through resistor R4. Since these two resistances are in series, the same current will flow through both of them. At some point, the current through R4 will place enough voltage across the resistor to turn on transistor Q4.

Note that the collector of Q4 is connected to the Q2/Q3/ZD1

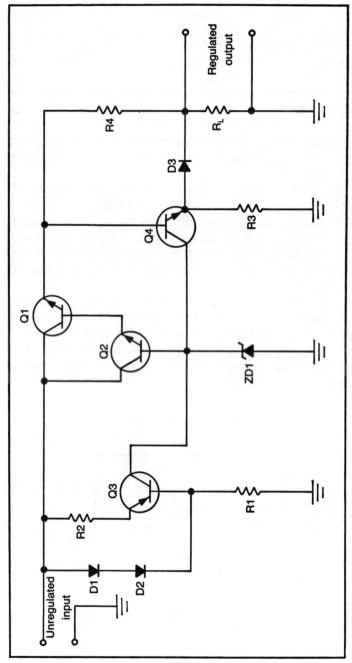

Fig. 8-20. Transistor Q4 provides output protection against shorts.

351

junction. When Q4 is conducting, it draws most of the current from Q3, so there isn't enough left to fully turn on the base/emitter junctions of the Darlington pair. As a result, Q1's collector current is reduced, preventing it from trying to pass an excessive amount of power and destroying itself.

At first glance, the circuit of Fig. 8-20 may have looked terribly complicated, but by examining it stage by stage, as in this discussion, we can see that nothing very complex is really going on.

Parallel Regulator. Another approach to voltage regulation with transistors is the parallel regulator, which is shown in its basic form in Fig. 8-21.

Current flows through resistor R1, and the zener diode to the base/emitter junction of the transistor. The voltage at the collector (across the load resistance (R_L)) is equal to the base/emitter junction voltage (0.7 volt) plus the breakdown voltage of the zener diode. A constant current source, and output short protection can be added to the circuitry, in the same manner as the series regulator discussed earlier.

Another form of the parallel regulator is illustrated in Fig. 8-22. Here current flows through R1, R3, the base/emitter junction of the transistor, and the zener diode. A fixed voltage is developed between the collector and the emitter of the transistor. The regulated output voltage can be found by using the following formula:

$$V_r = V_z \frac{R3 + R4}{R4}$$

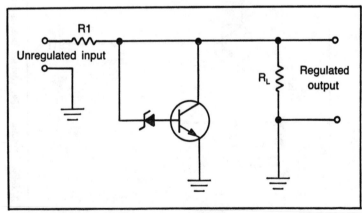

Fig. 8-21. A basic parallel regulator circuit.

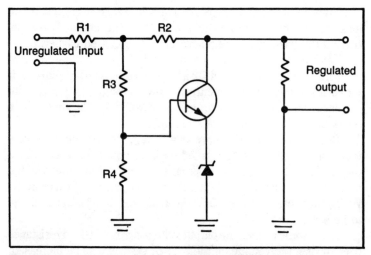

Fig. 8-22. A common variation on the basic parallel regulator circuit.

where V_z is the breakdown voltage of the zener diode.

As an example, let's consider a circuit in which the zener breakdown voltage is 6.8 volts, R3 is 3.9K (3900 ohms), and R4 is 4.7K (4700 ohms). In this case, the regulated output voltage would be equal to:

$$V_r = 6.8 \times \frac{3900 + 4700}{4700}$$

$$= 6.8 \times \frac{8600}{4700}$$

$$= 6.8 \times 1.83$$

$$= 12.44 \text{ volts}$$

The regulated output voltage from this type of circuit is always significantly greater than the zener diode's breakdown voltage.

The value of resistor R2 is rather critical, but it must usually be selected by trial and error to give the minimum variation in the voltage across R_L as the unregulated input voltage is adjusted from its minimum to its maximum value.

Feedback Regulators. Back in Chapter 3, you learned that feedback can often be used to improve the performance of transistor amplifiers. The same principles hold true for transistor voltage regulators.

Figure 8-23 shows a typical feedback regulator circuit. If you look closely, you should be able to see that this circuit is a variation on the basic series regulator circuit discussed earlier.

Let's examine how this circuit works.

First, transistors Q1 and Q2 are connected as a Darlington pair for series regulation. Input current flows through resistor R2 into the base Q2, and the collector of Q3. Meanwhile, the zener diode holds the emitter of Q3 at a fixed voltage.

Resistors R3 and R4 serve as a voltage divider, defining how much of the output voltage is fed back into the base of Q3. Since the R3/R4 combination is in parallel with R_L, the same voltage must be across both. The ratio between the two resistances in the voltage divider, therefore, allow the actual regulated output voltage to be set.

If, for some reason, the output voltage across the load resistance tries to rise, the voltage on the base of Q3 will also be increased. As a result, the transistor will start conducting more heavily. Since Q3 is now drawing more current from R2, there is less current available to reach the base of Q2. Current through the Darlington pair drops, so the current fed into R_L is reduced. Reducing the

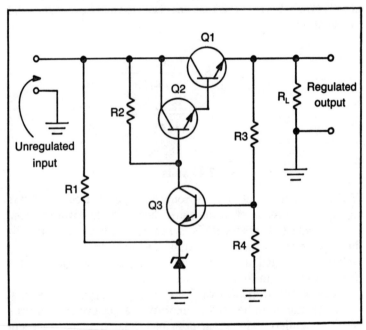

Fig. 8-23. A typical feedback regulator circuit.

current, will result in a lower voltage across the load resistance. The feedback circuit corrects for the output error.

The same process works in reverse if the regulated output voltage should try to drop for any reason. Less voltage is applied to the base of Q3, reducing that transistor's current flow, which increases the current flow through Q2 and Q1. The increased current through R_L creates a higher output voltage.

A feedback regulator circuit offers very precise voltage regulation, because the feedback process makes the circuit self-correcting. If the output value starts to drift in either direction, even by a small amount, the current flow through the regulator circuit will immediately readjust itself to compensate for the output error.

SWITCHING POWER SUPPLIES

All the voltage regulation circuits discussed so far in this chapter are linear circuits. They share some disadvantages which may or may not be of significance in a specific application.

By definition, a linear voltage regulator circuit must dissipate a fairly hefty amount of power. This means that large, heavy-duty (and expensive) power transistors must be used. In most cases, fairly extensive heat sinking will be required to prevent thermal damage to the semiconductors. Bulky transformers and large filter capacitors are also required. This all adds up to a large, heavy power supply circuit.

The input voltage must be greater than the output voltage. In addition, the output voltage of the regulator will always have the same polarity as the unregulated input signal.

Linear power supply circuits usually have only fair efficiency. The efficiency of a power supply circuit can be estimated by comparing the output voltage with the input voltage:

$$e = \frac{V_o}{V_i}$$

where e is the efficiency factor, V_o is the regulated output voltage, and V_i is the input voltage.

For instance, a regulated 9-volt power supply typically might require a 12.6 volt transformer to supply the input voltage:

$$e = \frac{9}{12.6} = 71.4$$

That isn't bad, but it's certainly not great either.

But is there an alternative? For years there wasn't, but in recent years the growing digital electronics technology has spawned the switching power supply. This type of circuit employs transistor switches to store energy in an inductor and a capacitor, to be supplied to the load circuit, as needed. Since the transistors are used as switches, they function in a nonlinear fashion.

Since the power transistor is either completely cut off, or in saturation (except for the brief transition times between states), the voltage is applied across the inductor, rather than across the transistor, as in a linear circuit. A transistor in a linear voltage regulator dissipates power, but an inductor in a switching power supply doesn't. As a result, less power is wasted, and the switching power supply offers very high efficiency, under all input and output conditions. Switching power supplies typically have efficiency factors of about 90%. This is clearly a considerable improvement.

Another big advantage of the switching power supply is that the output voltage can be greater than the input voltage (although, more current is consumed, of course, since you can't get more power out than you put in). This is a step-up regulator, since the voltage is stepped up.

Of course, a switching power supply can also be used as a step-down regulator, in which the output voltage is lower than the input voltage. With linear power supply and voltage regulation circuits, this is the only available choice.

A switching power supply circuit can also function as a voltage inverter, in which the output has the opposite polarity as the input signal. In a linear regulator, if the input voltage is positive with respect to ground, then the output must also be positive with respect to ground. But a positive input voltage to a switching power supply can result in either a positive or a negative output voltage depending on the specific circuit design used.

A switching power supply is more complex and requires a higher parts count than a comparable linear power supply/voltage regulator, but the components used can be considerably smaller and less expensive than the ones required in the linear circuit. A switching power supply can be very compact and lightweight, making it an excellent choice for portable equipment.

A switching regulator can be driven with a very poorly filtered dc signal. It can even be driven directly from three-phase rectifiers without any filtering in high power applications.

Still another major advantage of the switching regulator is its load-transient properties. A sudden momentary change in the load current results in only a very small, momentary change in the output voltage before the circuit self-corrects. The correction time is typically less than a few hundred microseconds (millionths of second).

It's not all gravy, however. Switching power supplies are subject to a number of significant disadvantages which have prevented the linear circuits from becoming completely obsolete.

The most obvious disadvantage is the increased circuit complexity. Linear power supply and voltage regulation circuits are generally straightforward and conceptually simple. Switching power supply circuits are more tricky.

The switching regulator's response time in correcting a sudden major change in the current drawn by the load circuit is not as good as many linear voltage regulator circuits.

The output of a switching regulator usually has more ripple than the output of a linear voltage regulator. This is because the switching operations in the circuit can contribute to ripple. A switching regulator can also add noise to the output signal. Noise can be reduced by filtering the input signal, reducing the series impedance, or increasing the switching time.

Faster response to sudden large changes in load current can be obtained by using low inductor values and high capacitor values. Another method is to sacrifice some efficiency, by keeping the $V_i:V_o$ ratio relatively large.

Let's consider the basic operating principles of the switching power supply. As stated earlier, the transistor(s) is used as an electronic switch. This switch is turned on and off at a rate determined by the input voltage, the output voltage, and the load current to supply the load with the desired power.

The control circuitry adjusts the switching frequency in response to these factors. Obviously a feedback loop is required for the control circuitry to sense the output voltage.

A relatively large output capacitor stores the energy when the transistor is cut off to provide an average flow of current through the load resistance.

Energy is also stored in an inductor during the transistor's cutoff time. This also maintains the current flow to the load. Current flow return is through a diode.

The equivalent series resistance of the capacitor at the switching frequency (about 20 kHz) is extremely important. Even when high quality (low leakage) capacitors are used, higher capacitance values than would be used in a comparable linear circuit must be employed in the switching regulator.

A typical step-down switching regulator circuit is shown in Fig. 8-24. The output voltage will be less than the input voltage. This circuit is occasionally referred to in technical literature as a forward converter.

The coil (inductor) is in series with the output load. Energy is applied simultaneously to the load and the inductor.

When the transistor is saturated (on), the voltage at the emitter rises to the input voltage minus the saturation voltage of the transistor. The difference between the saturated emitter voltage and the output voltage is across the coil.

The load current flows from the transistor, through the inductor and the capacitor. When the output voltage increases beyond a specific level, sensed by the control circuitry through the feedback loops, the transistor will be cut off. Current will still continue to flow for a while after the transistor is switched off. Then the voltage at the emitter is dropped to the voltage drop of the diode (typically about 0.7 volt). This diode is in the inductor's current path. The voltage across the coil is minus the sum of the diode's voltage drop and the output voltage. As long as the transistor is off, the inductor current will decrease towards zero. When the output voltage drops below a specific level, the control circuitry will

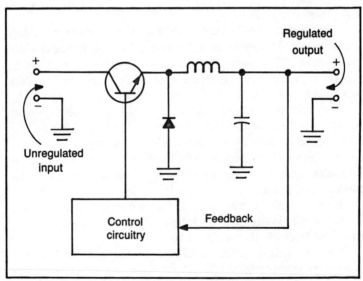

Fig. 8-24. A typical step-down switching regulator circuit.

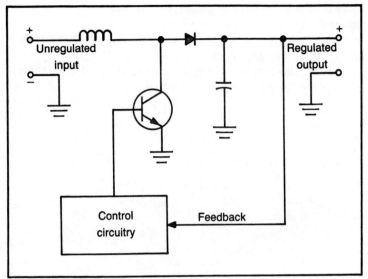

Fig. 8-25. A typical step-up switching regulator circuit.

turn the transistor switch back on again for another cycle.

The duty cycle of the transistor switching signal is selected so the average current through the coil is equal to the load current. The capacitor's average current is zero. The result of these factors is a constant output voltage. The larger the value of the

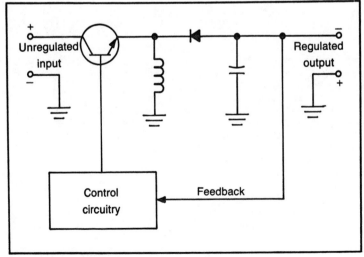

Fig. 8-26. A typical inverting switching regulator circuit.

capacitor, the smoother the output voltage will be (i.e., the less ripple).

Figure 8-25 shows a typical step-up switching regulator circuit, and a typical inverting switching regulator circuit is illustrated in Fig. 8-26. Notice that the same components are used in all three circuits, only their positions are changed.

Chapter 9

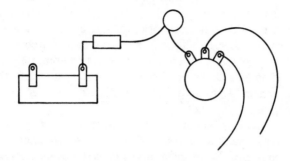

General Design Tips

We have barely scratched the surface in describing the potential applications of the bipolar transistor. We have covered the most important, and commonly used types of applications, and you should now feel you have a good head start on figuring out what is happening in almost any transistor circuit you might come across. If a circuit looks terribly complicated, mentally break it into stages. Most complex circuits are made of simpler subcircuits.

By the same token, in designing your own circuits, you will usually be combining the simple basic circuits described in this book (or variations on them) in various ways.

In this chapter we will give a few general tips the circuit designer should keep in mind. They are especially significant for complex (many stage) designs.

CONSTRUCTION METHODS

In most cases, you will probably build your first prototype circuit on some kind of solderless breadboarding system, as described in the experiments in the earlier chapters of this book. This is an extremely good idea. As we will see a little later in this chapter, a design that looks good on paper won't always work the way we expect it to. Often some experimentation is required to get the exact results we want from our design. A solderless breadboarding system makes changes in component values and circuit wiring very simple and convenient to accomplish.

Remember to always make changes with the power off to reduce the risk of damaging some components, or, in some circuits, suffering severe injury yourself. Remember at all times that electrical shocks can be very painful, or even fatal—always use caution.

Some circuits may not work well (or at all) in a standard solderless breadboard. This is usually true in high frequency circuits where the length of connecting wires, phantom components, and shielding (or lack thereof) become critical. Fortunately, the circuit that cannot be prototyped in a breadboard socket is very much the exception to the rule.

Somewhat more common are circuits which may change their operating parameters somewhat when a more permanent type of connection is used. Be aware of this potential problem. There usually isn't very much you can do about it in advance, but at least you'll be able to recognize such problems when they do crop up. This can, at least, save a lot of time and hair pulling.

Once you have designed your circuit and breadboarded the prototype, and gotten all the bugs out, you will probably want to rebuild it in a more permanent way. Solderless breadboarding sockets are great for testing and experimenting with prototype circuits, but they aren't much good when it comes to putting the circuit to practical use.

Breadboarded circuits, by definition have nonpermanent connections. In actual use, component leads may bend and touch each other, creating potentially harmful shorts. Components can even fall out of the socket altogether. Interference signals (discussed shortly) can easily be generated or picked up by the exposed wires.

Generally, packaging in a circuit built on a solderless breadboarding socket can be tricky at best. They just don't fit all that well in standard circuit housings and boxes. Besides, a solderless breadboarding socket is fairly expensive. It is certainly worth the price if it is re-used for many circuits, but if you tie it up with a single permanent circuit, you're only cheating yourself. Less expensive methods that are more reliable, more compact, and that offer better overall performance are readily available.

Some, relatively simple circuits can be constructed on a perforated circuit board, with component leads and jumper wires soldered directly together with point to point wiring. Only very small, very simple circuits should be wired directly together without any supporting board. Otherwise you will end up with a "rat's nest" of wiring that is next to impossible to trace if an error is made or if the circuit needs to be serviced at a later date. In addition, a lot of loose wires can create their own problems, such as stray capacitances and inductances between them allowing signals to get into the wrong portion of the circuits.

"Rat's nest" wiring is also begging for internal breaks within the wires, and short circuits. Momentary, intermittent shorts may not cause permanent damage in all cases, but they can cause strange circuit performance that can be maddeningly frustrating to diagnose and service.

Rat's nest wiring can also be a problem with complex circuits on perf boards. Try to minimize crossings of jumper wires. Use straightline paths for jumper wires whenever possible. In fact, it is a good idea to minimize the use of jumper wires. Some will be unavoidable, of course, but many can be eliminated by repositioning the components. Schematic diagrams are usually drawn to avoid too many lines crossing (which would make the diagram difficult to read). They give you a starting point in laying out the components for a minimum of jumper wires. You can rarely follow the arrangement of components in the schematic exactly (and it often is undesirable to do so, even if it were physically possible), but the diagram can get you started in the right direction.

Try to position all the components on the board to see how they'll fit before you even plug in your soldering iron. This will help you avoid unpleasant surprises, like ending up with no place to put that big filter capacitor.

For moderate to complex circuits, or for circuits from which a number of duplicates will be built, a printed circuit board gives very good results. Copper traces on one (or in very complex circuits) on both sides of the board act as connecting wires between the components. Very steady, stable, and sturdy connections can be made, since the component leads are soldered directly to the board itself. Great care must be taken in designing the layout of a PC board to eliminate wire crossings (traces cannot cross each other. If a wire crossing is absolutely essential, a wire jumper must be used), and to minimize stray capacitances between traces that could affect circuit performance. In critical cases, a guard band between traces can help reduce the potential problem. This is shown in Fig. 9-1.

The copper traces are usually placed very close to each other. This means a short circuit is very easy to create. A small speck of solder, or a piece of component's excess lead could easily bridge across two adjacent traces, creating a short.

Tiny, near invisible hairline cracks in the copper traces can also

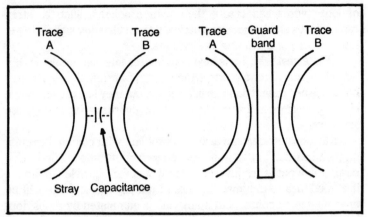

Fig. 9-1. A guard band can be added to minimize phantom capacitances between adjacent copper traces on a PC board.

be problematic if you're not careful. Generally, fairly wide traces that are widely spaced are the easiest to work with. However, they aren't always practical (or even possible) for many circuits, especially circuits using integrated circuits.

A printed circuit board type of construction results in very short component leads. This can help minimize interference and stray capacitance problems.

Care must be taken when soldering any semiconductor component. Too much heat can permanently damage the delicate crystaline material. Transistor sockets are available, but they add to the cost of building the circuit, and don't provide as firm a connection as soldering the transistor's leads directly into the circuit. I would only recommend using a transistor socket when a very expensive device is being used (maybe), or in an application where the transistor will frequently be taken in and out of the circuit, or replaced often. For a transistor tester, for instance, a transistor socket would obviously be desirable.

Transistors usually have a very long life-span, so there is no need to use sockets to make servicing more convenient, as is the case with vacuum tubes.

To minimize the risk of overheating when soldering a transistor (or diode, or other heat-sensitive component), the lead being soldered should be heat sinked with a paper clip, a pair of needle-nosed pliers, or a commercially available spring loaded clamp made for just that purpose.

More and more circuits being constructed today use the wire-wrapping method. A thin wire is wrapped tightly around a square post. The edges bite into the wire, making a good electrical connection without soldering. Components are fitted into special sockets that connect their leads to the square wrapping posts. This form of construction is really only suitable for circuits made up primarily of a number of integrated circuits. If just a few discrete components are used, they can be fitted into special sockets, or soldered directly, while the connections to the ICs are wire-wrapped. In circuits involving many discrete components (including most transistor circuits), the wire-wrapping method of construction is not very practical.

For most of the circuits designed with the help of this book, printed circuit boards, or perforated boards are usually the best choice.

CONTROL PANELS

One aspect of design that too few designers pay enough attention is to control panel layout. All too often controls are positioned however they'll physically fit.

Some consideration should be given to those aspects of design which are sometimes referred to as "human engineering." The control panel should be arranged for the human operator's convenience.

Controls should not be packed too tightly together. Leave room for someone with large fingers to get a good grasp on each con-

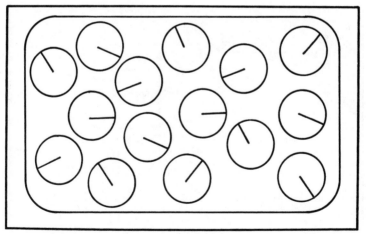

Fig. 9-2. Staggering controls allows you to fit more controls in a small space without packing them too tightly.

trol. If a great many controls are needed in a relatively small space, maximum use can be made of that space by staggering the controls, as illustrated in Fig. 9-2.

Always use good quality switches and potentiometers. Using budget controls is false economy. They wear out and become unreliable to quickly. You can often get good prices on high quality switches and potentiometers from surplus dealers.

Controls that are held in place with washers and nuts, are far preferable to ones that must be glued, or otherwise permanently fixed in place. The potentiometer you glue to the control panel may need to be replaced at some later date. Don't make things harder on yourself, or on others who might be using your equipment.

All controls should be clearly and unambiguously marked. Try to avoid abbreviations that might be misinterpreted. A label that reads VOL, could probably be reasonably assumed to mean "VOLume," but what about a label that says "AMP"? Does it mean "AMPeres," or "AMPlification," or "Atomic Monopole Processor"?

In designing a control panel, try to keep controls with related functions grouped together. This is especially important for equipment with a lot of controls. If the operator has to search the entire panel for the specific control he needs, the control panel design is inefficient.

Try to limit the number of wires from the control panel to the main circuit board, especially if the board is not mounted directly

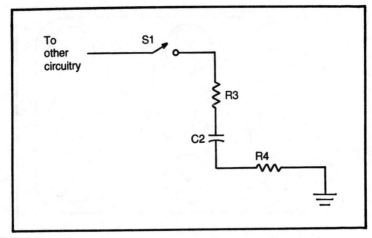

Fig. 9-3. This partial circuit should probably be mounted directly on the control panel.

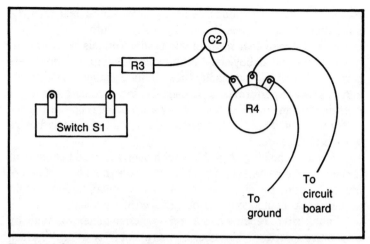

Fig. 9-4. A few parts can be mounted right on the lugs of the controls.

on the control panel itself. The more connecting wires there are, the more chance there is for one of them to develop a break.

Some components can be mounted directly on the lugs of the controls. For example, consider the partial circuit shown in Fig. 9-3. It would be silly to run a wire from switch S1 back to the circuit board for resistor R3 and capacitor C2, then run another wire back from the circuit board to potentiometer R4. It makes more sense to mount resistor R3 and capacitor C2 directly on the lugs of S1 and R4, as shown in Fig. 9-4.

If there is a common line that is connected to several of the controls, only run one wire back to the circuit board, and connect the appropriate control lugs together in daisy chain fashion, as illustrated in Fig. 9-5. This is often done with ground and Vcc connections, but it may come up for signal connections too.

COMPACTNESS

A small, lightweight package is certainly desirable for many applications. Don't waste space. To use a 5 1/2" × 7" × 3 1/4" box for a circuit that would fit into a 2" × 3 1/2" × 2" box with room to spare is simply wasteful.

On the other hand, many designers (including many designers of commercial equipment) are going to the other extreme lately. They go for compactness at the expense of other factors.

Some circuits (especially those carrying relatively high power signals) can get quite warm. It is a good idea to leave breathing

367

room for heat dissipation. Use a somewhat larger box than the minimum the circuit board will physically fit into.

It is assumed that most of the readers of this book are experimenters and hobbyists. Many experimenters and hobbyists periodically update, and modify their circuits, adding new features, and improving performance. If there is even a remote chance of ever making any modifications of any kind, leave room for the circuitry to grow. This practice can save you a lot of grief and extra work and expense.

Probably the biggest problem with overly compact equipment is the lack of convenient serviceability. Leave room for test equipment probes on the circuit board. Also leave enough room so that an individual component can be replaced if necessary.

I once had to repair a commercial function generator made by a major manufacturer. The housing was about 6″ × 4″ × 3″, which was about the minimum practical size for the controls. When I opened the case, I found that 30-gauge wires were used to connect the circuit board to the controls. Several of the wires when the case was opened, and several others broke during the course of the repair. Of course, this meant a lot more extra work for me.

Never use light gauge wire for off-board connections. There is no need to go overboard and use 16-gauge wire. I'd recommend

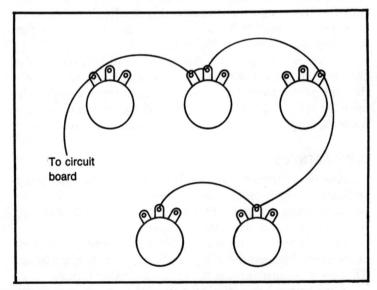

Fig. 9-5. Common connections to multiple controls should be made at the control panel.

something in the 24-gauge to 20-gauge range, for most applications. If the equipment is likely to be jolted around a lot, use a heavier gauge wire. Using 30-gauge wire for off-board connections (especially in a piece of test equipment supposedly designed for field use) goes far beyond the realm of stupidity.

Inside the case of this function generator was a tightly packed 1″ × 1 1/2″ pc board, and nothing else. The components were so tightly packed, I could barely get a pair of tweezers in to grasp the component that needed replacement. This was completely inexcusable. There was plenty of room in that case for a larger circuit board. If a 2″ × 3″ board had been used the components would have been close enough to prevent interference pickup by the connecting leads, but spaced enough to be serviceable, and to have better heat dissipation (there was an IC in there that ran very hot).

If this function generator had been competently designed, the repair would have taken about half an hour. As it was, it took over two hours, and a lot of heavy cursing.

The electronics industry has gotten into a frame of mind that states that "smaller is always better." Much of modern electronics technology has grown out of the space program, and it has brought some habits along with it. In a rocket, all circuitry should be as lightweight and compact as possible, or more expensive fuel will be needed to get it into space. Serviceability is not an issue, since the equipment itself will probably be inaccessible.

In consumer equipment here on Earth, however, we need to make a compromise between compactness and serviceability. Don't make extra work for yourself (or another technician)—especially when it doesn't give any practical advantage (beyond the possible aesthetic satisfaction of cramming a complex circuit into a record-breaking small space).

INTERFERENCE

It is always a good idea to keep all connecting wires as short as possible. A wire can act as an antenna. It may start transmitting noisy signals that could interfere with the operation of other circuits (or even of other stages in the same piece of equipment). On the other hand, it may pick up other stray signals that can interfere badly with the signals that are supposed to be in the circuit. The most common culprits for inducing interference into a circuit are radio frequency signals and broadcasts, and the low frequency hum of the ac power lines.

Interference problems can be reduced by shielding the circuit. Any circuit that might be sensitive to stray signal pickup should be enclosed in a metallic housing. A metal case should also be used with any circuit that uses high frequency signals (including most digital circuits) to prevent the circuit from acting as an unintendional radio transmitter.

Interconnecting wires that are susceptible to interference pickup or transmission can be replaced with shielded cable.

Another technique, which is especially good for limiting low frequency pickup is to use twisted pairs of wires. Two signal leads are twisted together between their source and destination point.

DERATING COMPONENTS

Many people who work with electronics, either professionally, or as a hobby, feel that if a component is not operated outside its rated values, it should last forever, unless some outside force (dropping the equipment, a short circuit, or whatever) damages it. Regrettably, this is not the case.

Electronic components do wear out with time. This is why electronics servicing is a multibillion dollar industry. It is up to the designer to stack the odds against component failure for the longest possible period.

Of course, no component should be operated outside its nominal maximum limits, as defined by the manufacturer. For example, if a capacitor is rated for a maximum working voltage of 100 volts, do not apply 120 volts. Exceeding the maximum ratings will greatly reduce the expected life span of the component. In many cases, even momentarily exceeding the rated limit can destroy the component almost instantly.

If we have a capacitor with a maximum working voltage rating of 100 volts, can we use it in a circuit that will apply 95 volts to it? Well, yes, we could, but it would not be a very good idea. For one thing, we may get an unexpected overvoltage. Perhaps component tolerances in another part of the circuit, could turn the nominal 95 volts into 105 volts. Or perhaps a momentary short or other circuit defect may affect the voltage across the capacitor. This sort of thing happens all the time. If we don't leave enough "elbow room" in the component ratings, it may very soon be a case of "bye bye, capacitor."

But, even without the possibility (almost probability) of unexpected increases in the circuit's operating parameters, it is still not a good idea to operate a component too close to its maximum rated level.

To see why this is so, we have to introduce a factor called *electrical stress*, which is the ratio of the applied voltage to the maximum rated voltage:

$$s = \frac{V_a}{V_r}$$

where s is the electrical stress factor, V_a is the applied voltage, and V_r is the rated maximum voltage. For higher values of electrical stress, the *failure rate* (probability of failure within a given period of time) increases exponentially. A relatively small change in the electrical stress factor can result in quite a large change in the failure rate.

Temperature is also an important factor in a component's lifespan probability. The higher the ambient or surface temperature, the higher the failure rate.

Temperature and electrical stress interact. If one must be increased for any reason, the other should be decreased to keep the same failure rate.

The failure rate of a component is a measure (usually in percent) of how likely it is to survive for a specified period of time, without failure. Another way of looking at it is the percentage of identical parts that will fail within that time. For example, if one thousand identical transistors are operated under exactly the same conditions for 500 hours, and 15 fail, the failure rate for 500 hours is 0.015, or 1.5%.

Bear in mind that one or two of those failures may have occurred very early in the test—perhaps, within the first hour or two. The individual component you are using may fail at any time. You can only play the percentages, and gamble that is one of the more durable majority.

You can improve the odds by not operating the component close to its limits. For example, in the thousand transistors mentioned above, operating at a somewhat lower voltage may result in a failure rate of 0.008, or 0.8%.

The designer should select a derating factor, and multiply all specified limits for the components by that derating factor.

As an example, let's say we want to use capacitors rated for 60 volts maximum, and 1/2-watt resistors. If we use a derating factor of 75%, what are the maximum values to be applied to these ponents? For the capacitors, the applied voltage should not exceed:

$$V_{ar} = DF \times V_r$$

(where V_{ar} is the adjusted maximum voltage, DF is the derating factor, and V_r is the component's maximum voltage from the manufacturer's specification sheet.):

$$V_{ar} = 0.75 \times 60 = 45 \text{ volts}$$

No more than 45 volts should ever be applied to any of the capacitors under any normal operating condition.

Similarly, for the resistors, the wattage should never be allowed to exceed:

$$W_{ar} = DF \times W_r = 0.75 \times 0.5 = 0.375 \text{ watt}$$

The maximum ambient and surface temperatures should be similarly derated. It is a good idea to use extensive heat sinking whenever there is any chance that the temperature might reach a significant level (noticeably above room temperature).

It might seem that the best approach would be to use the heaviest duty components available. For example, using a 2500 volt capacitor to filter a 9-volt power supply should reduce the electrical stress factor to an almost negligible value:

$$s = \frac{9}{2500} = 0.0036 = 0.36\%$$

This may not give any significant advantage over using a 100-volt capacitor. Below a certain point, the failure rate becomes almost a constant. There is nothing in particular to be gained by decreasing the electrical stress factor further. Doing so will actually introduce certain disadvantages. The higher rated component will be larger, heavier, and more expensive than what is needed.

In addition, some components should not be operated too much below their rated levels. For example, electrolytic capacitors tend to dry out if they are left unused on the shelf too long. The electrical stress (in a sense) rejuvenates the electrolyte. Operating the electrolytic capacitor at a voltage much below its rating will be the same as leaving it on the shelf—it will eventually dry out, and fail.

As a rule of thumb, stick to derating factors in the 50% to 85%

range. Especially pay attention to the maximum ratings for semiconductors. There are often several ratings that must be dealt with concurrently.

COMPONENT TOLERANCES

The rated value of a component is rarely exactly its true value. The manufacturer only guarantees the approximate value. A tolerance rating is usually given. The actual value will be between $-x\%$ and $+x\%$ of the nominal value. The tolerance rating is usually written as $\pm x\%$.

We will consider resistors to illustrate component tolerances. Other components will work in the same basic way.

Most resistors have one of three standard tolerance ratings:

5%
10%
20%

Precision resistors with tolerances of 1% 0.5%, or even 0.1% are also available.

Let's assume we have a bunch of 10K resistors. How close can we expect their actual values to be to their nominal values? For a 5% tolerance resistor, the actual value may range from a low of:

$$R_L = 95\% \text{ of } 10,000 = 0.95 \times 10,000 = 9500 \text{ ohms}$$

to a high of:

$$R_H = 105\% \text{ of } 10,000 = 1.05 \times 10,000 = 10,500 \text{ ohms}$$

Five percent tolerance resistors have become the unspoken standard of modern electronics. Modern digital electronics require greater precision than circuits using older technologies.

The former standard was the 10% tolerance resistor. For a unit with a nominal value of 10,000 ohms, the actual value may be as low as:

$$R_L = 90\% \text{ of } 10,000 = 0.90 \times 10000 = 9000 \text{ ohm}$$

or as high as:

$$R_H = 110\% \text{ of } 10,000 = 1.10 \times 10000 = 11000 \text{ ohms}$$

Twenty percent tolerance resistors should only be used in very noncritical applications, since their actual value can vary so much from the nominal value:

$$R_L = 80\% \text{ of } 10{,}000 = 0.80 \times 10000 = 8000 \text{ ohms}$$

$$R_H = 120\% \text{ of } 10{,}000 = 1.20 \times 10000 = 12{,}000 \text{ ohms}$$

Bear in mind that even a 20% tolerance resistor may be exactly at the nominal value, but the manufacturer only guarantees that it will be within ±20% of the stated value.

Some hobbyists buy surplus high tolerance resistors, then measure their exact values with an ohmmeter. Once the exact value is known, it won't change unless the resistor is damaged in some way. Unfortunately, measuring resistors this way is exceedingly dull and tedious at best. If you've got the time and inclination, by all means give it a try. Personally, I think buying 10% or 5% tolerance resistors in the first place is well worth the few extra pennies they cost. For most applications, they will be close enough, even at the extremes of their tolerance.

In some applications, very precise values are required. There are several things you can do here. One is to measure individual resistors until you find an actual value that is to close enough to the required value. Another method is to use a small trimmer potentiometer in series with another resistor of slightly less than the required value. The trimmer potentiometer is adjusted until the precise value is obtained. This method is especially handy when oddball, nonstandard values are called for.

The final possibility is to buy a precision (low tolerance) resistor of the correct value. These devices are significantly more expensive, but if that is what is needed for the circuit to work properly, they are certainly worth the price.

Capacitors generally have wider tolerance ratings than resistors. It is more difficult to manufacture capacitors with exact ratings, so high precision (low tolerance) capacitors tend to be rather hard to come by, and quite expensive when you do locate them.

Electrolytic capacitors have extremely wide tolerances—often as much as ±50%. Fortunately, these components are rarely used in applications where the exact value is terribly critical.

If just one or two components are not quite at their nominal value, most designs won't be affected too severely. But most practical circuits consist of dozens of components. The odds are that the cumulative error will more or less cancel itself out. That is, some components will be somewhat below their nominal value, and others will be above their nominal value.

The designer, however, should give some consideration to the worst possible case. It is always possible that all the errors will be in the same direction, and the cumulative effect can be very severe. Don't scoff—it can and does happen. Sometimes a design that looks perfect on paper won't work when it is actually built. Assuming there are no errors in the wiring, the culprit is undoubtably component tolerances.

Consider the simple resistive voltage divider network circuit shown in Fig. 9-6. For the resistance values shown on the diagram, the three output voltages should have the following values:

$$A = 6.28 \text{ V}$$
$$B = 5.18 \text{ V}$$
$$C = 3.56 \text{ V}$$

The nominal values in this circuit are summarized in Table 9-1a. In Table 9-1b we look at the actual values that might be found in the circuit if typical ±5% tolerance resistors are used. The deviation percentages for the individual resistors are arbitrarily selected. Another set of ±5% resistors would give somewhat different

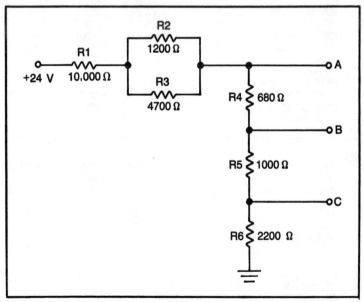

Fig. 9-6. Simple resistance voltage divider circuit used to demonstrate the effects of component tolerences.

**Table 9-1. (A) Nominal Values for the
Circuit of Fig. 9-6. (B) Typical Results When Using 5%
Tolerance Resistors. (C) Typical Results using 20% Tolerance Resistors.**

Actual	% Off from Nominal Value	Actual Value	Voltage Drop
A			
R1	0	10,000	16.18
R2	0	1,200	- -
R3	0	4,700	- -
R4	0	680	1.10
R5	0	1000	1.62
R6	0	2200	3.56
R2/R3 combination	- -	956	1.55
total circuit resistance		14836	24.01
CURRENT		1.618 mA	
OUTPUT A		6.28 V	
OUTPUT B		5.18 V	
OUTPUT C		3.56V	
B			
R1	+3.3%	10330	16.39
R2	−1%	1188	- -
R3	−5%	4465	- -
R4	+4.2%	708.56	1.12
R5	+1.2%	1012	1.60
R6	−2.7%	2140.6	3.40
R2/R3 combination		938.34	1.49
total circuit resistance		15,129.5	23.992
CURRENT		1.59 mA	
OUTPUT A		6.12V	
OUTPUT B		5.00 V	
OUTPUT C		3.40 V	
C			
R1	+8.3%	10830	16.46
R2	+5.5%	1266	- -
R3	−0.7%	4667	- -
R4	−10%	612	0.93
R5	+3.8%	1065	1.62
R6	+6.5%	2284	3.47
R2/R3 combination		996	1.51
total circuit resistance		15,787	23.99
CURRENT		1.52mA	
OUTPUT A		6.02 V	
OUTPUT B		5.09 V	
OUTPUT C		3.47 V	

Actual	% Off from Nominal Value	Actual Value	Voltage Drop
R1	–18%	8200	15.16
R2	+11%	1332	- -
R3	–3%	4559	- -
R4	+17%	796	1.47
R5	+4%	1040	1.92
R6	–13%	1914	3.54
R2/R3 combination		1031	1.91
total circuit resistance		12,981	24
CURRENT		1.85 mA	
OUTPUT A		6.93 V	
OUTPUT B		5.46 V	
OUTPUT C		3.54 V	

results. Look at what the cumulative tolerance error did to the output voltages:

$$A = 6.12 \text{ V} = 97\% \text{ of nominal value}$$
$$B = 5.00 \text{ V} = 96\% \text{ of nominal value}$$
$$C = 3.40 \text{ V} = 95\% \text{ of nominal value}$$

Table 9-1c gives us the results for a typical collection of ± 10% tolerance resistors. The output voltages in this particular instance are:

$$A = 6.02 \text{ V} = 97\% \text{ of nominal value}$$
$$B = 5.09 \text{ V} = 98\% \text{ of nominal value}$$
$$C = 3.47 \text{ V} = 97\% \text{ of nominal value}$$

As it worked out, the ± 10% resistors didn't do much worse than the ± 5% resistors. In fact, for outputs B and C, they even did slightly better. Remember, the tolerance only identifies the maximum deviation from the nominal value. The actual value may be close to the nominal value, but there is no guarantee.

Similarly, in a circuit with a number of components, the positive deviations may cancel out the negative deviations, as in Table 9-1c, but they may not. The designer must be prepared for the worst possible error.

Table 9-1d summarizes the circuit values when a typical collection of ± 20% tolerance resistors are used. In this case we get the following output voltages:

$$A = 6.93 \text{ V} = 110\% \text{ of nominal value}$$
$$B = 5.46 \text{ V} = 105\% \text{ of nominal value}$$
$$C = 3.54 \text{ V} = 99\% \text{ of nominal value}$$

Component tolerances can introduce quite a bit of variation into a circuit.

Because of component tolerances, sometimes a replacement component will not give the desired results, even though it is nominally equivalent to the original.

Especially be careful with *general replacement* transistors. Usually they will work in the same circuits as the type numbers they are considered replacements for, but the specifications may not be exactly the same. A general replacement is defined as a unit that is approximately similar to the original. A general replacement transistor may not work in certain critical applications.

When in doubt, compare the specification sheets for both the general purpose transistor you want to use, and the original transistor you want to use, and the original transistor you want to replace. Make sure that the important specifications are close enough for your intended application.

THE MOST IMPORTANT TIP

We will close this chapter with probably the most important and labor-saving tip of all. Far too many designers do not follow this practice, and they inevitably end up making more work and frustration for themselves.

What is this important tip? *Make a lot of notes and detailed diagrams throughout the design process. Don't count on your memory— Write it down!!!*

This is especially important for complex circuits, but you'll be surprised at how often you end up having to redo some calculations and layouts even with a simple circuit if you don't keep good notes.

The notes don't have to be grammarical. Misspellings, and poor penmanship are OK, as long as you can read the notes when you need them. Illegible notes not only make more work; they can result in errors that can set your design back quite a few steps. You don't want to mistake a 7 for a 1, or a 4 for a 9. Especially be careful with decimal points. Make them large, dark, and solid, so there is no chance of mistaking a speck of dust for a decimal point, or vice versa.

By the same token, you don't have to draw up neat, publishable quality diagrams. The only thing important here is that you clearly record the relevant information. It doesn't matter if you can't draw a straight line, as long as you can tell what the lines represent.

Label everything in all diagrams. That's so important, I think I'll repeat it, with emphasis—*Label everything in all diagrams.* Don't rely on your memory to determine what that little squiggle is— label it, so there is no possibility of mistaking what it is supposed to be. Even label what seems perfectly obvious. It only takes a few seconds, and if you ever need that label (and you undoubtably will, more often than you might expect), you will be extremely grateful that it's there.

Chapter 10

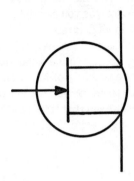

Other Transistor Types

Throughout this book we have concentrated primarily on standard bipolar transistors. A bipolar transistor, you should recall, is made of three slabs of semiconductor material, appropriately doped, and arranged in "sandwich" fashion, as shown in Fig. 10-1.

There are many other types of transistors, besides just the bipolar. In this final chapter we will briefly examine a few of the other most important transistor types.

THE UJT

The bipolar transistor has two internal PN junctions, as suggested by the "bi-" prefix in its name. Another important transistor type is the UniJunction Transistor, or UJT. The name implies there is only a single PN junction.

The internal structure of an N type UJT is illustrated in Fig. 10-2.

The schematic symbol for an N type UJT is shown in Fig. 10-3.

As you probably have already guessed, there is also a P type UJT, in which the N type semiconductor slab and the P type semiconductor slab are reversed. Figure 10-4 shows the schematic symbol for a P type UJT. For the rest of our discussion, we will concentrate on the N type device.

Notice there are three leads—an emitter (the P type section in an N type UJT), and two base connections on either end of the larger N type section.

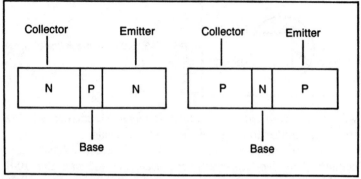

Fig. 10-1. Bipolar transistors made up of "semiconductor sandwiches."

Essentially, the N type section acts as a voltage divider resistor pair, with a diode (the PN junction) connected to the common ends of the two resistances. A simplified equivalent circuit for an N type UJT is shown in Fig. 10-5.

An important specification of a UJT is the *intrinsic standoff ratio*. This value is determined by the internal base resistances:

$$\eta = \frac{R_a}{R_a + R_b}$$

η is the symbol used to represent the intrinsic standoff ratio. It typically has a value somewhere around 0.5, because the values of R_a and R_b are usually fairly close to each other.

If a voltage is applied between base 1 and base 2, the diode

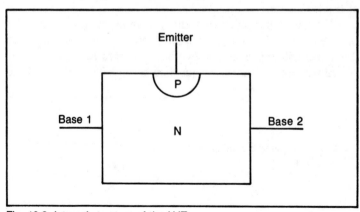

Fig. 10-2. Internal structure of the UJT.

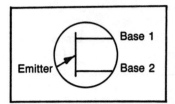

Fig. 10-3. Schematic symbol of an N-type UJT.

will be reverse-biased. Naturally, this means no current will flow from the emitter to either of the base terminals.

Now, let's suppose there is an additional variable voltage source connected between the emitter and base 1, as shown in Fig. 10-6. As this emitter voltage is increased, eventually a point will be reached where the diode becomes forward-biased. Beyond this point, current can flow from the current to the two base terminals.

The normal voltage at the junction is equal to the product of the voltage applied across the two bases times the intrinsic standoff ratio. That is:

$$V_j = \eta \, V_{bb}$$

In the circuit of Fig. 10-6, the output pulses are generally tapped off across resistor R2. R1's function is to keep the circuit operating properly despite fluctuations in temperature. An additional, complementary (180° out of phase) output may be tapped off across this resistor.

The values of these two resistors are usually significantly smaller than the internal resistances of the UJT itself. This allows us to ignore the external resistances when analyzing the operation of the transistor.

When an input pulse with enough amplitude to forward-bias the PN junction in the UJT is applied, output pulses (in step with the input) will appear at the two bases (across R2 and R1).

At what point does the PN junction become forward-biased and

Fig. 10-4. Schematic symbol for a P-type UJT.

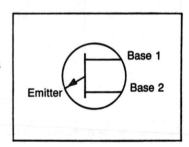

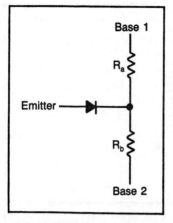

Fig. 10-5. The equivalent circuit for a UJT.

start to conduct? This point can be approximately defined with the following formula:

$$V_c = 0.5 + V_j = 0.5 + \eta V_{bb}$$

where V_c is the conduction voltage, V_j is the junction voltage, and V_{bb} is the voltage applied across the bases. When V_e is equal to or greater than V_c, the UJT will conduct.

Adding an RC network to the emitter circuit, as shown in Fig. 10-7, creates a simple relaxation oscillator.

When the supply voltage (V_{bb}) is first applied to the circuit,

Fig. 10-6. The basic UJT circuit for demonstration purposes.

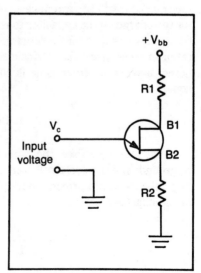

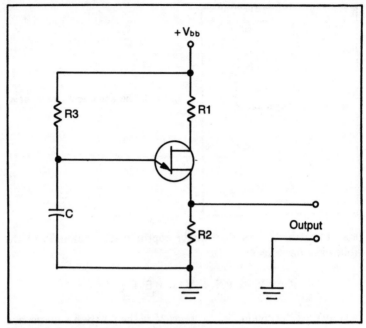

Fig. 10-7. A simple UJT relaxation oscillator circuit.

the voltage across the capacitor (C) is obviously 0 volts. The voltage across this component will start to increase as the capacitor charges. At some point, the voltage across C will exceed the V_c point, and the transistor will fire (produce an output pulse). The time it takes for the charge on the capacitor to reach this point is dependent on the values of capacitor C, resistor R3, and the intrinsic standoff ratio. In most cases, the intrinsic standoff ratio can simply be ignored, and the charging time is considered to be approximately equal to:

$$T_c \cong CR3$$

When the UJT fires, the capacitor will quickly be discharged to ground, and the cycle will start all over. The discharge time is negligible, so the approximate frequency is equal to the reciprocal of the charging time:

$$F = \frac{1}{T_c} \cong \frac{1}{CR3}$$

In this case it is best to initially select a resistor value, and then solve for the necessary value of C:

$$C = \frac{1}{FR3}$$

This is because, ideally, the resistance should be between:

$$\frac{V_{bb}(1-\eta) - 0.5}{2I_p}$$

and:

$$\frac{2(V_{bb} - V_v)}{I_v}$$

where I_p is the maximum current flowing from the emitter to base 1, V_v is the "valley voltage" (the emitter/base 1 voltage just after the UJT has started to conduct), and I_v is the "valley current" (the emitter/base 1 current when V_v is across the junction.

Most of these values can be found in the manufacturer's specification sheet for the device. Typical values are listed. The actual value may vary somewhat around the nominal value.

Let's start a typical design example. Our goal is to design a UJT relaxation oscillator with an output frequency of 750 Hz, at 2.5 volts peak-to-peak (across R2).

The UJT we will be using in our design has the following specifications:

$$\eta = 0.52$$
$$I_p = 10 \ \mu A = 0.00001 \ \text{amp}$$
$$V_v = 3.1 \ \text{volt}$$
$$I_v = 25 \ \text{mA} = 0.025$$
$$\text{Internal resistance } (R_a + R_b) = 9000 \ \text{ohms}$$

The first step is to select the supply voltage (V_{bb}). We will design our circuit to operate on 12 volts.

Now we need to find the acceptable range for R3. This resistance should have a value between:

$$\frac{V_{bb}(1 - \eta) - 0.5}{2I_p} = \frac{12(1 - 0.52) - 0.5}{2 \times 0.00001}$$

$$= \frac{12(0.48) - 0.5}{0.00001} = \frac{5.76 - 0.5}{0.00002}$$

$$= \frac{5.71}{0.00002} = 285,500 \text{ ohms}$$

and:

$$\frac{2(V_{bb} - V_v)}{I_v} = \frac{2(12 - 3.1)}{0.025} = \frac{2 \times 8.9}{0.025}$$

$$= \frac{17.8}{0.025} = 712 \text{ ohms}$$

As you can see, we have quite a range of acceptable values for R3. Let's try 22K (22,000 ohms). This would mean capacitor C would be equal to:

$$C = \frac{1}{FR3} = \frac{1}{750 \times 22000} = \frac{1}{16500000}$$

$$\cong 0.000000061 \text{ farad} = 0.061 \ \mu F$$

This is not a standard capacitance value. We might be able to find a 0.062 μF capacitor, but we will probably have to settle for a 0.047 μF unit. This changes the output frequency to:

$$F = \frac{1}{CR3} = \frac{1}{0.000000047 \times 22000}$$

$$= \frac{1}{0.001034} = 967 \text{ Hz}$$

That isn't very close to our target value of 750 Hz.

Since we have found a capacitance value that almost fits, we can go back and calculate a new value for R3 that should still be within the desired range:

$$R3 = \frac{1}{FC} = \frac{1}{750 \times 0.000000047}$$

$$= \frac{1}{0.0000353} = 28{,}369 \text{ ohms}$$

The nearest standard resistance value is 27K (27,000 ohms). This is still well within the acceptance range (712 ohms to 285,500 ohms). This rounding off of the resistance value gives us an output frequency of:

$$F = \frac{1}{CR3} = \frac{1}{0.000000047 \times 27000}$$

$$= \frac{1}{0.001269} = 788 \text{ Hz}$$

This will be close enough for most applications. After all, component tolerances might account for as much error (see Chapter 9). If a more precise output frequency is required, try placing a 2.5K (2,500 ohms) trimpot in series with R3. This will allow the actual resistance to vary from 27,000 to 29,000 ohms. Since the nominal ideal value for this resistance is 28,369 ohms, we have plenty of "elbow room" on either side to also compensate for component tolerances.

Now, what about the values of R1 and R2? The equation for finding the desired value of R1 is:

$$R1 = \frac{R_a + R_b}{2 \, \eta \, V_{bb}}$$

For our example design, this works out to a resistance of:

$$R1 = \frac{9000}{2 \times 0.52 \times 12} = \frac{9000}{12.48}$$

$$= 721 \text{ ohms}$$

R1 will usually have a value from about 500 ohms to 1000 ohms. This resistance value is not extremely critical, and we can round off to the nearest standard value available. In this case, we can use either a 750 ohm or a 680 ohm resistor.

Now, we need to find a value for R2. The formula is:

$$R2 = \frac{((R_a + R_b) + R1)V_o}{V_{bb} - V_o}$$

where V_o is the peak-to-peak output voltage across R2.

For our design example, resistor R2 should have a value of approximately:

$$R2 = \frac{(9000 + 680)2.5}{12 - 2.5} = \frac{9680 \times 2.5}{9.5}$$

$$= \frac{24200}{9.5} = 2547 \text{ ohms}$$

We can use either a 2.2K (2,200 ohms) or a 2.7K (2,700 ohms) resistor for R2.

THE PUT

A close relative of the UJT is the PUT, or Programmable Unijunction Transistor. The basic structure of this device is illustrated in Fig. 10-8. The schematic symbol for a PUT is shown in Fig. 10-9.

The three leads of a PUT are labeled as follows:

> A (Anode)
> C (Cathode)
> G (Gate)

A voltage is placed across the anode and the cathode, with the anode positive with respect to the cathode. No current will flow

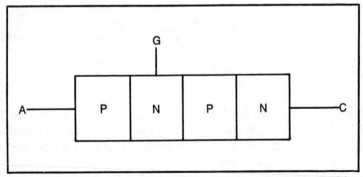

Fig. 10-8. The internal structure of a PUT consists of four semiconductor slabs.

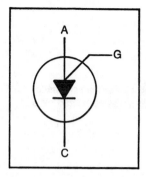

Fig. 10-9. The schematic symbol for a PUT.

between the anode and cathode until a negative (with respect to the anode) pulse is applied to the gate.

The unique feature of the PUT is that its intrinsic standoff ratio is not predetermined by the internal characteristics of the PUT itself. Instead, the value of η can be set anywhere between 0 and 1, by selecting the proper values for external resistances.

THE SCR

A SCR, or Silicon Controlled Rectifier, is a semiconductor device that is very similar to a PUT. The structure of a SCR is shown in Fig. 10-10. Compare this sketch with the structure of the PUT, shown in Fig. 10-8. The primary difference is that, in the SCR, the gate is near the cathode, rather than near the anode.

The schematic symbol for an SCR is shown in Fig. 10-11. Again, compare this with the symbol for the PUT, which was shown in Fig. 10-9.

The leads of an SCR are labeled in the same way as for a PUT (Anode, Cathode, and Gate).

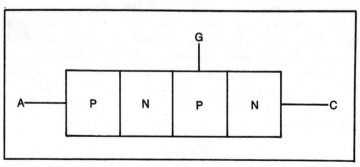

Fig. 10-10. The main difference between an SCR and a PUT (Fig. 10-8) is the placement of the gate.

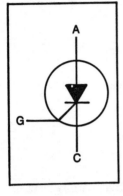

Fig. 10-11. Notice how similar the schematic symbol for a SCR is to the symbol used for a PUT (Fig. 10-9).

An SCR is essentially an electrically switchable diode. If a voltage is applied between the anode (+) and the cathode (-), and the gate is grounded (0 volts), no current will flow, even though the pseudo-diode is forward-biased.

If an increasingly positive voltage is applied to the gate, it will eventually reach a specific trigger voltage (which is dependent on the specific SCR used). Now, current can flow from the cathode to the anode, as if through an ordinary forward-biased diode.

If the gate voltage is not removed, current will continue to flow through the device. This current will continue to flow until the voltage between the anode and the cathode is interrupted.

THE FET

Many transistor circuits are variations on older vacuum tube circuits. There are many similarities in operation between transistors and vacuum tubes. However, the operation of a standard bipolar transistor doesn't quite correspond to that of a vacuum tube. Some circuits won't work as well as if vacuum tubes were used.

Does this mean we have to forego modern semiconductor technology, and revert to bulky, expensive, and hot vacuum tubes? Not at all. A Field Effect Transistor, or FET, is a semiconductor device that can closely mimic the operation of a vacuum tube. In addition, it is capable of several unique tricks of its own.

The basic internal structure of a FET is illustrated in Fig. 10-12. Notice that there are three leads, which are labeled:

S—(Source)
D—(Drain)
G—(Gate)

These names will be explained shortly.

The body of an N type FET is a single, continuous length of N type semiconductor material. A small section of P type material is placed on either side of the N type section. Both P sections are electrically tied together. Their common lead is the gate.

P type FETs, in which the types of semiconductor material are reversed, are also available.

To get a general idea of how a FET works, consider the mechanical water system illustrated in Fig. 10-13. When the valve (gate) is opened, as in Fig. 10-13A, the water can flow through the pipe, from its source, to where it can drain out.

If, on the other hand, the valve is partially closed, as in Fig. 10-13B, the amount of water that can flow through the pipe is limited. Less water comes out of the drain.

If the valve is completely closed off, no water at all will be able to flow through the pipe. Nothing will come out of the drain.

In the same way, the gate terminal of a FET controls the amount of electrical current that can flow from the source to the drain.

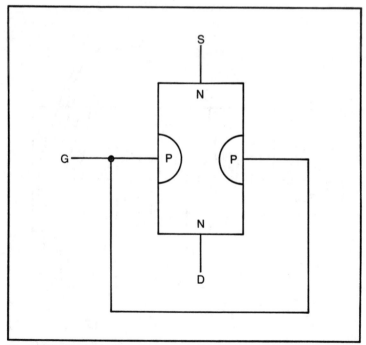

Fig. 10-12. The basic internal structure of a FET.

A negative voltage applied to the gate reverse-biases the PN junction, producing an electrostatic field (electrically charged region) within the N type material slab. This electrostatic field opposes the flow of electrons through the N type section, acting somewhat like the partially closed valve in our mechanical model. The higher the negative voltage applied to the gate, the less current that is allowed to pass through the device from the source to the drain.

This is directly analogous to the action of a vacuum tube. The voltage on the grid controls how much current flows from the cathode to the anode. The gate of a FET, of course, corresponds to the grid, the source to the cathode, and the anode to the anode. An FET is, in a very real sense, a solid-state vacuum tube.

Like a vacuum tube, a FET has a very high input impedance, so it will draw very little current. Since, according to Ohm's Law:

$$I = \frac{E}{Z}$$

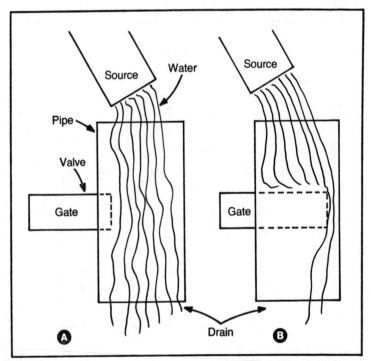

Fig. 10-13. A FET works something like a valve closing off a water pipe.

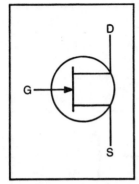

Fig. 10-14. The schematic symbol for an N-type FET.

increasing the impedance (Z), will decrease the current (I), for a given voltage (E).

This high input impedance means that FETs can be used in highly sensitive measurement applications, and in circuits where it is important to avoid loading down (drawing heavy currents from) previous circuit stages.

The schematic symbol for an N type FET is shown in Fig. 10-14. Correct biasing is illustrated in Fig. 10-15.

The current path from the source to the drain is sometimes called the channel.

THE MOSFET

The field effect transistors discussed in the last section are more properly called Junction Field Effect Transistors, or JFETs. They are not the only type of Field Effect Transistor available.

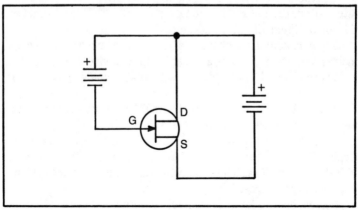

Fig. 10-15. Simple circuit illustrating the correct biasing of an N-type FET.

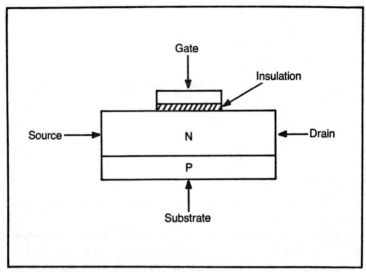

Fig. 10-16. The gate section of a MOSFET is physically isolated from the semiconductor channel.

Another type does not have an actual PN junction at all. These devices are known generically as Insulated Gate Field Effect Transistors, or IGFETs. As the name clearly states, the gate is insulated from the channel.

The most common way of doing this is by using a thin slice of metal as the gate (rather than a slab of semiconductor crystal). This metal is oxidized on the side that is placed against the semiconductor channel. This insulates the gate from the semiconductor, because metal oxide is a very poor conductor. When a metal oxide gate is used, the IGFET is often called a MOSFET, or Metal Oxide Silicon Field Effect Transistor.

The semiconductor channel is backed by a substrate of the opposite type of semiconductor. That is, if the device is built around an N type channel, the substrate will be a P type semiconductor slab.

The basic internal structure of a MOSFET is illustrated in Fig. 10-16. The schematic symbols are given in Fig. 10-17.

Even though the gate is physically insulated from the channel, it can still induce an electrostatic field into the semiconductor to limit the current flow. The substrate is generally kept at the same voltage as the source.

Correct biasing for an N channel MOSFET is illustrated in Fig. 10-18.

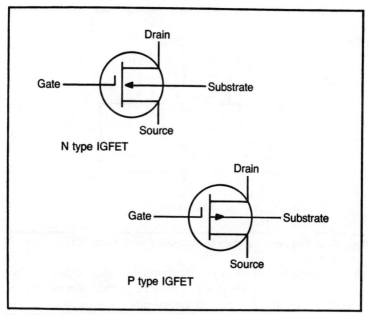

Fig. 10-17. The schematic symbols for IGFETs.

THE ENHANCEMENT MODE FET

So far we have being talking only about depletion mode FETs. These devices reduce current flow by increasing the negative voltage on the gate (assuming an N type channel). Another type of FET operates in the enhancement mode. In this type of FET the source and drain are not parts of a continuous piece of semicon-

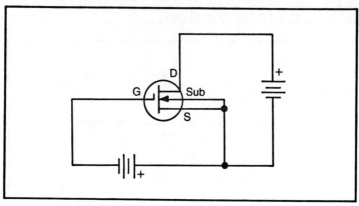

Fig. 10-18. Simple circuit illustrating the correct biasing for an N channel MOSFET.

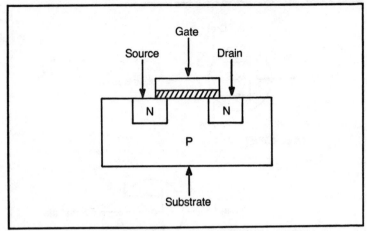

Fig. 10-19. The basic structure of an enhancement mode FET.

ductor material. They are separated from each other by the substrate. The basic internal structure of an enhancement mode FET (with an N type channel) is illustrated in Fig. 10-19.

A positive voltage is normally applied between the gate and the source. The higher this voltage is, the greater the number of holes drawn from the N type source into the P type substrate. These holes are then drawn into the N type drain region by the voltage applied between the drain and the source.

In other words, increasing the voltage on the gate increases the amount of current flow from the source to the drain.

The schematic symbol for an N channel enhancement mode FET is shown in Fig. 10-20. All enhancement mode FETs are of the insulated gate type (IGFETs).

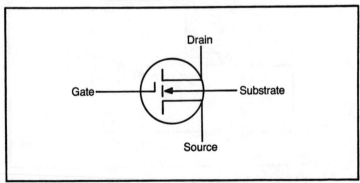

Fig. 10-20. the schematic symbol for a N-type enhancement mode FET.

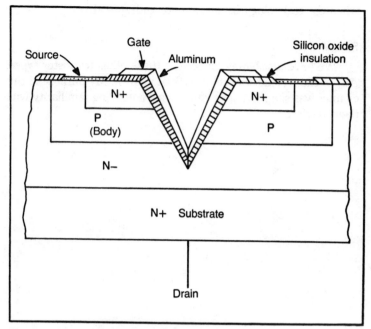

Fig. 10-21. The internal structure of a VMOS.

THE VMOS

In the last few years, a new type of "Super-FET" has been developed. Its basic structure is illustrated in Fig. 10-21. This is the VMOS, or Vertical Metal Oxide Semiconductor device.

A VMOS offers a considerable improvement over regular MOS devices. This is due to several factors, including greater current density, and higher drain/source breakdown voltage.

Some of the advantages of VMOS devices include:

> Very high input impedance
> Very rapid switching (in the nanosecond range)
> Low feedback capacitance
> High transconductance
> Linear transfer characteristics
> No thermal runaway
> Low on-state voltage
> No offset voltage

VMOS devices are still very new, and so they are rather expensive, but the prices are expected to come down in the near future when the technology becomes more widespread.

SUMMARY

We have just barely glanced at these other transistor types. Any of them could warrant a book of its own. My purpose in this chapter was simply to give you at least a nodding familiarity with these devices.

Appendices

Appendix A
Transistor Projects

METER AMPLIFIER

Of all dc milliammeters ever manufactured, the one having a full-scale deflection of one milliampere (sensitivity, 1,000 ohms-per-volt) was probably the most common. At one time this meter was a workhorse but with improvements in magnets and manufacturing techniques, much more sensitive meters are being used. The great big hitch here, of course, is that the more sensitive the meter, the more it costs. The circuit shown in Fig. A-1 will give a meter an apparent ten-time increase in sensitivity, but by the time you get through paying for the parts, you may not have made any savings at all. But at least its nice to know that you have a choice. And there's a limit to how sensitive a meter instrument can be made! A circuit such as this is thus very useful when we need a super-sensitive meter.

To understand the circuit of Fig. A-1, think of the current amplification capability of a transistor. In an earlier chapter we called this beta. It is the ratio of the change in collector current to a change in base current. For a representative transistor, beta could have a value, say, of 50. Thus, if we had 20 microamperes of base current, we would have 1,000 microamperes (or a milliampere) of collector current. And if the base current changes, we get 50 times as much change in the collector current.

Now suppose that in the circuit of Fig. A-1 we had a dc milliammeter in the base circuit of the transistor. We would need a

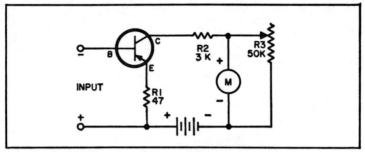

Fig. A-1. Insensitive meter can be used to measure currents in the order of microamperes with the help of a transistor amplifier.

very sensitive meter indeed to give us a full-scale deflection of 20 microamperes. But if we had a dc meter in the collector circuit, we could easily use a 1 milliampere movement. We could calibrate this meter in microamperes and when it would read full-scale we would be measuring a current of only 20 microamperes.

Is the circuit new? Not actually, since it is a very ordinary, very simple transistor amplifier. What we have here is a dc amplifier with but one new component. This is potentiometer R3, a calibration adjustment that is varied until the meter reads full-scale.

It is important to understand just what it is that the meter in the collector circuit does. It just reads collector current. But this collector current is the same as the base current multiplied by the beta of the transistor. Thus we can take advantage of the amplifying action of the transistor to measure small currents. Remember, though, that these small currents are not measured directly, but indirectly.

RELAY AMPLIFIER

Just as the circuit in Fig. A-1 permits us to use a less sensitive meter (and therefore a less expensive meter) so too can we use this circuit, or some variation of it, to give us an apparent increase in the sensitivity of other components. A relay is one such item. Relays, like meters, are assessed in terms of their sensitivities.

Fig. A-2 shows the relay circuit. The relay might require a current of one or more milliamperes to attract the armature of the relay toward its core. In the circuit of Fig. A-2 a current of about 50 microamperes in the input will increase the collector current to the amount required to close the relay. Assuming this current to be 2 milliamperes (or 2,000 microamperes), the gain in our circuit is 40 (40 × 50 = 2,000). In this circuit the relay is the load.

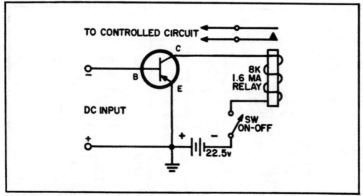

Fig. A-2. Simple transistor amplifier can be used to operate a relay.

In the circuits of Figs. A-1 and A-2, what would be our input? It could be some sort of signal which would give us the necessary amount of forward bias to activate the collector circuit. The input current could also be produced by a solar cell, as shown in Fig. A-3. Now take a few moments out to note the considerable similarity these circuits have.

PREAMPLIFIER

At first glance, the circuit in Fig. A-4 looks far more complicated than any of those we have shown so far in this chapter. For this reason we have drawn a dashed line down the center of the circuit so that you can see for yourself that all we have here is just a pair of audio amplifiers, with the first amplifier driving the second.

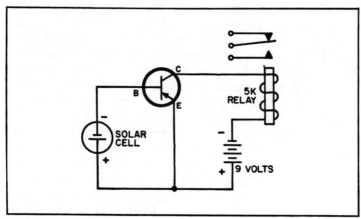

Fig. A-3. Relay circuit will operate when solar cell receives light.

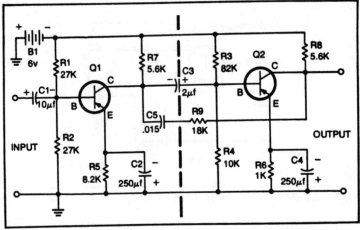

Fig. A-4. Resistance-coupled audio amplifier consists of a pair of amplifiers, one driving the other.

We have analyzed the components in circuits similar to this one in earlier chapters, but let's take out a few brief moments to do it again, just by way of review. If you will examine R1 and R2 you will see at once that these two resistors are in series. If you will trace their connections to battery B1 you will see that they are connected directly across it. These two resistors supply the correct amount of forward bias for Q1. Now what about R3 and R4? Exactly the same as R1 and R2! They supply bias for Q2.

Note that R1 and R2 do not have the same values as R3 and R4. If Q1 and Q2 are different you would expect their forward biasing requirements to be different also. The two transistors also use emitter bias. The emitter resistors, R5 and R6 are shunted by large values of capacitance—250 μF. These capacitors have a rating of 6 volts. While we're on the subject of capacitors, examine C1 and C3. These are also fairly large capacitance units (compared to the types used in vacuum-tube amplifiers) and are 25-volt electrolytics. We must watch our polarity in the case of electrolytics. When considering the emitter bypass capacitors, remember that for PNP transistors the electrons flow from the emitter down to ground. This makes the top end of the emitter resistor minus and the bottom end plus. C2 and C4 are connected to observe this polarity.

In the case of C2 and C4 the matter of polarity is quite clear cut, but what do we do about coupling capacitors C1 and C3? It is entirely possible for a coupling unit to be connected to two positive voltage points, yet we must still be careful about polarity.

To see how we can come to a decision about this, please look at Fig. A-5. In Fig. A-5A we have an electrolytic directly across a battery. There is no problem here since the plus terminal of the electrolytic goes directly to the plus terminal of the battery. Similarly, the negative lead of the electrolytic is wired to the negative end of the battery. The electrolytic receives the full voltage of the battery and must be able to withstand it.

In Fig. A-5B, we are still observing polarity since the negative end of the electrolytic is connected to the arm of the potentiometer. But isn't point B a positive point? Yes, it is, but it is less positive than point A.

To simplify matters even further, we have eliminated the potentiometer in Fig. A-5C. Now you can see that we have connected the negative terminal of the capacitor to one of the positive terminals of the battery. This is point B. But so is point A. However, while point B is positive, it is less positive than point A. Another way of saying exactly the same thing is to say that point B is negative with respect to point A.

Now getting back to Fig. A-4, how do we know that we have connected C1 and C3 correctly, so that polarity is observed? In the case of C1 we don't know. We have not been told just what it is that C1 connects to. It is possible that in connecting C1 to its input we would find it necessary to turn C1 around.

We have no such doubts about C3. The collector of Q1 to which this capacitor is connected is more negative than the base of Q2, to which the capacitor is also connected.

R7 and R8 are easy to identify. These are our collector load resistors. What is left? Just a series combination of C5 and R9. This

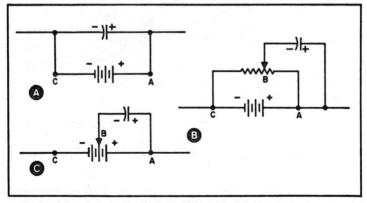

Fig. A-5. Methods of connecting an electrolytic capacitor.

is our negative feedback circuit. It improves the stability of the amplifier and gives the amplifier a somewhat better frequency response.

We have gone through this amplifier rather hurriedly, since audio amplifiers were covered in an earlier chapter. A simple resistance-capacitance coupled amplifier of this sort could be used as an audio preamplifier.

SIGNAL TRACER

A signal tracer is just an audio amplifier plus a detector. The detector is generally a crystal diode as shown in Fig. A-6. For the amplifier portion we could have used the circuit of Fig. A-4, but we have added the one in Fig. A-6 just to emphasize the fact that not all audio amplifiers are exactly alike. As an example, locate C3 (Fig. A-6). If you will trace its connections you will see that it is shunted directly across the battery. The purpose of this capacitor is to maintain a low-impedance path around the battery. As long as the battery is fresh, or at least not too used, its impedance will be fairly low. But with time and use the internal resistance of the battery will rise, and long before a decision is made to discard it, will act as a common impedance or coupling device between stages. This can result in feedback between stages, producing amplifier instability. An electrolytic across the battery ensures a low-impedance path around it regardless of the condition of the battery. But before we decide that this is a wonderful cure all, remember that electrolytics also get old and that they have their own ills—including an increase in their own internal impedance.

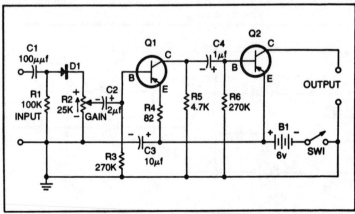

Fig. A-6. The diode permits use of the audio amplifier as a signal tracer.

Now examine the coupling capacitors in Figs. A-4 and A-6. How do they compare as far as capacitance and polarity are concerned? C3 in Fig. A-4 has the same capacitance and the same positioning as coupling capacitors C2 and C4 in Fig. A-6. But what about the input capacitors (C1 in Fig. A-4 and C2 in Fig. A-6)? Quite a difference here in several respects. Note the arrow in Fig. A-6 showing the direction of current flow through the diode load resistor R2. When the slide arm is at the bottom end of the resistor the negative end of C2 is connected to the negative end of the battery. But suppose the slide arm is at the top end of the potentiometer. The lead of C2 is now at some plus point depending on the amount of current being rectified by the diode and the value of the gain control. This might be as much as 1 or 2 volts. Now what about the other side of C2? This side of the capacitor is connected to the base of Q1. From the base we have two ways of approaching the battery. One path is through a fairly high value of resistance (R3—270,000 ohms) to the negative terminal of the battery. The other path is through the transistor to the emitter, through 82-ohm emitter of R4 to the plus terminal of the battery. This latter path is so very much lower in resistance than R3 that the base (hence capacitor C2) is not too far removed (in terms of voltage) from the plus terminal of the battery.

Since this may not be too clear from an examination of the circuit, we have re-drawn it for you in Fig. A-7. Consider the emitter and base of the transistor as a resistor of low-value in series with R4 and R3. These three components act as a voltage divider across the battery. But because of the very large value of R3, the plus connection of C2 is much closer, from a voltage viewpoint, to the plus terminal of the battery.

While we are making comparisons between the two amplifiers, is there anything else we should notice? Yes—there are several items we should spot. Q2 in Fig. A-6 has no emitter resistor. The emitter resistor for Q1 isn't bypassed. Also, the two transistors, Q1 and Q2 act as part of a voltage dividing network. In Fig. A-4 however, we had an actual resistive voltage divider network (R1 and R2; R3 and R4) for forward bias. This arrangement is better than compelling the transistor to work in a resistive function. Also, in Fig A-6 we have no feedback network. From all this you might conclude that Fig. A-4 is the better circuit, and so it is. But before we toss out Fig. A-6, consider the uses of the two circuits. One is an audio preamplifier where we have some interest in stability and sound quality. The other is intended as a rather ordinary audio

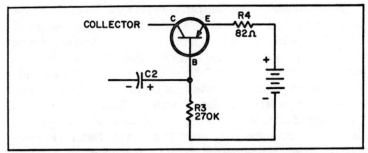

Fig. A-7. Rearrangement of part of the circuit of Fig. A-6 to show voltage-divider action.

amplifier to be used as part of a test instrument.

It is possible that the difference between these two circuits may have been immediately apparent to you. Carry this sort of analysis along to transistor radio receivers and you will at least have learned a few of the reasons why there is a price differential among them.

AUDIO AMPLIFIER

As long as we're examining audio amplifiers, let's take a look at the one shown in Fig. A-8. This is a transformer-coupled unit, but a first glance seems to show some component arrangements worth investigating. The emitter circuit for Q1 looks a bit odd until we examine it more closely and see that some of the components have been drawn in such a way as to make the circuit look peculiar. R4 and C3 we can recognize immediately as the emitter-resistor and bypass capacitor. But what about R2, R3, and C5. We can't do much about these until we grope around a bit and locate R8. Look just a bit

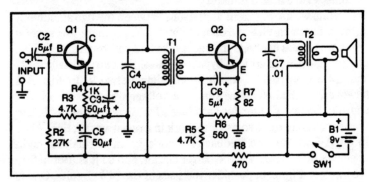

Fig. A-8. Sometimes radio components are drawn so that their function isn't too obvious. When this happens start tracing the circuit and all the parts will fall into their proper places.

more carefully and you will see that R2, R3, and R8 are all in series and that this series network is connected directly across the battery. What is this network? It supplies forward bias for Q1. Similarly, R5, R6, and R8 have the job of forward biasing Q2.

What other information can we extract from this circuit? C4 in the output circuit of Q1 might require a bit of thinking. The small value of C4 indicates that it is an rf bypass, somewhat unusual to find in an audio amplifier. However, if the input is supplied through a diode detector, we might find C4 necessary. It is possible, however, that the diode detector might have an rf bypass in its own output circuit, so C4 might be an unnecessary luxury. C7, in the output of Q2 is a fixed tone-control capacitor.

TRANSISTOR RADIO

The circuits we have examined so far in this chapter are certainly modest enough in their demands for parts, yet you would have a complete transistor receiver with still fewer components. We have the circuit in Fig. A-9. And just in case you think that this is good for earphone reception only, we have included provisions for a speaker.

A receiver of this sort will not have the sensitivity or selectivity of a six-transistor set, nor is there any claim of this sort being made. What we are interested in here is not in running a contest between receivers but in learning more about the functioning of transistor circuits.

With this caution in mind, what is there about Fig. A-9 that can add to our storehouse of knowledge? You might think that the obvious absence of parts might clarify the situation.

The first item we might notice is that Q1 is being used as a detector. The fact that it is marked detector is a sure-fire giveaway. How could we know, then, that it is a detector without the help of the label? There are several clues that demand attention.

Q1 is located between a transistor (Q2) being used as an audio amplifier and a tuning coil. Even if Q2 were not marked, we see that it is connected to a phone jack. Consequently we can draw the conclusion that the input to Q2 must be audio. And since the input is a tuned rf circuit (it would have to be with a 365 $\mu\mu$F tuning capacitor) then the only function left to Q1 would be detection.

Suppose, though, we assume this isn't enough evidence. What more information do we have? What about the forward bias for Q1? Actually, there is no forward bias other than that supplied by the input signal. And since this is the case, we may assume that the

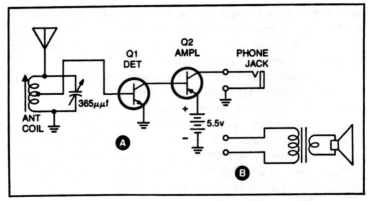

Fig. A-9. Transistor radio receiver with a minimum number of parts.

operating point on the characteristic curve must be in the vicinity of the cutoff point. In this circuit we depend on one-half of the input signal to supply enough forward bias to take the transistor out of cutoff. The other half of the signal will drive it further into the cutoff region. This is rectification, or, to give it its other name, detection.

What about Q2? Is it biased at all? Is it properly biased? Since we have a pnp transistor we want our emitter to be positive with respect to the base. We can see quite readily that the emitter connects to the plus terminal of the battery, but what about the base. The return path of the minus terminal of the battery is through the emitter-collector circuit of Q1. Before we decide this is not for us, consider the emitter-collector of Q1 as a resistive path and you'll have no trouble getting back to the negative terminal of the battery.

Working with crystalline materials, as we are, does not necessarily imply that all our explanations are going to be crystal clear.

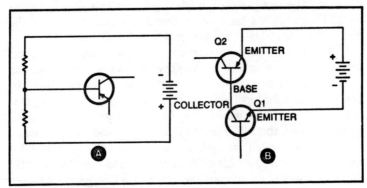

Fig. A-10. A pair of transistors can be used as a voltage divider.

Earlier in this chapter we mentioned that a circuit diagram will often obscure an understanding of the theory. For this reason let us consider the biasing of Q2 once again, but this time from the vantage point of Figs. A-10A and B. In drawing A the biasing arrangement for the transistor is the same circuit arrangement we have used so often in earlier chapters. But what about drawing B? We don't have to connect the base of Q2 into the divider since Q2 and Q1 in series act very much like the resistive pair shown in Fig. A-10A.

Appendix B

Power Supply Projects

5-VOLT DC, 1-AMP POWER SUPPLY

The 5-volt, 1-ampere power supply shown in Fig. B-1 is usable for a large variety of TTL and CMOS digital projects, as well as those analog projects that will operate from this low voltage. The basis for this power supply is a three-terminal IC voltage regulator. The 5-volt type of regulator was one of the earliest available in three-terminal IC form. The LM-309K is now considered the venerable predecessor for all three terminal IC voltge regulator devices. In this application, you can use either the LM-309K, LM-340K-5 or the 7805, all three are approximately equivalent devices.

When selecting a voltage regulator, keep in mind the current rating that you are building for. Not all seemingly equal devices are usable in all cases. The problem is the package. There are three types. The "H" package (LM-309H) is usable in free-air without heatsinking to only 100 milliamperes. The "H" package is the same as the small, metal "T0-5" transistor can. The "K" package will support up to 1 ampere unheatsinked and some drive it to 1.5 amperes with suitable heatsinking (not recommended). The "K" package (LM-309K) is the same as the T0-3 diamond-shaped power transistor case. The popular "T" package (LM-340T-5) is a plastic power transistor case and it will safely support only 750 milliamperes without heavy heatsinking. For this power supply the "K" package devices are most heartily recommended. If the "T" devices are used, derate the output current accordingly.

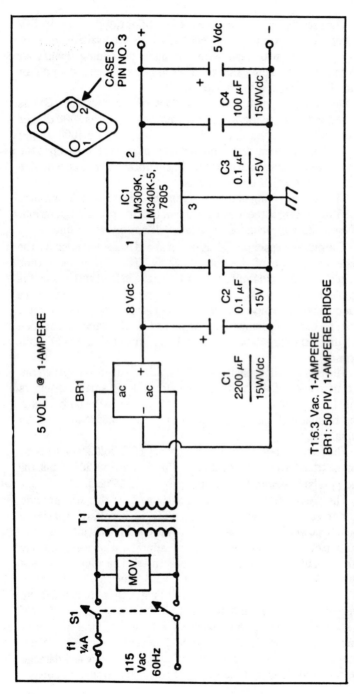

Fig. B-1. A 5-volt, 1-ampere power supply.

413

The rectifier is a full-wave bridge rectifier rated at 50 volts PIV at 1 ampere of current. These rectifier stacks are available in black epoxy packages with lead wires or pins, depending upon your skeleton. Do not mount this rectifier in a manner that cuts off air circulation or overheating could result.

The filter capacitor at the output of the rectifier (C1) is selected according to the "rule-of-thumb" that requires 2000 μF per ampere of load current. Some authorities only require 1000 μF per ampere, but I recommend the higher value as good practice up to the point where the additional capacitance significantly adds cost or size (not a factor in small current supplies).

The voltage rating of the capacitor should be 15WVdc or more. The 15 volt rating is the lowest that will provide reliable operation. A 25-volt, 35-volt or more capacitor will not be out of line.

The output capacitor, C4, is optional and is used to improve the transient response of the power supply. There is a short time required for the regulator to react to changes in load current so it is possible to experience a "suck-out" of supply votlage in the face of a sudden heavy demand for current. The charge stored in capacitor C4 can be used to "dump" into the circuit in the event of a sudden need until the regulator can catch up with the demand (milliseconds).

The two 0.1 μF capacitors (C2 and C3) are used to improve the noise immunity of the voltage regulator. These small capacitors should be located physically as close as possible to the body of pins of the regulator. Otherwise, the effect of these capacitors will be minimal.

The transformer used in this supply is a 6-3-volt ac@1-ampere filament transformer (filaments are used in vacuum tubes, but the name persists despite the fact that filaments do not).

The "MOV device" is a *metal oxide varistor* that will provide better rejection of power line noise. The high voltage transients normal to power lines in many parts of the country will interrupt the operation of electronics devices, especially digital devices, and may cause damage to the regulator and rectifier. The MOV is a two-terminal device that is made by General Electric. It is readily available in electronic parts stores. It is strictly optional and the power supply will work nicely without it. Do not delete the MOV if you know the power line is noisy or if there is a local history of strange happenings on electronic equipment in your area. When computers just "bomb" for no reason, and do not show any damage to be repaired, is a good indication of power line transient noise.

Make sure that the switch used will support 110 volts ac across the pins. The purpose of the double-pole-single-throw switch is to break both sides of the power line (a safety feature).

+12-VOLTS DC, 1-AMP POWER SUPPLY

Another power supply like the first (refer to the information on the 5-volt, 1-ampere power supply) is the +12-volt, 1-ampere power supply shown in Fig. B-2. This supply differs from the others in the ratings of its components. The transformer, for example, is a 12.6-volt ac, 1-ampere, "filament" transformer. Also, the rectifier has a high PIV voltage rating.

The integrated circuit voltage regulator (IC1) is a LM-340K-12 or a 7812 device. Again, the "K" package is preferred if a full 1 ampere of current is required. Otherwise, this power supply is the same as the other.

±12-VOLT DC, 1-AMP POWER SUPPLY

If we connect two single-polarity power supplies "back-to-back," we can make the form of dual-polarity power supply needed by operational amplifier circuits (and with other linear IC devices). But that method is wasteful of transformers and rectifiers and therefore is not well regarded.

The circuit shown in Fig. B-3 is a better solution that will require only one transformer and one rectifier. The transformer is a 25.6 volt ac *center-tapped* type rated at not less than 2-amperes. The popular Triad 2.8 volt or the Radio Shack 3-ampere models are preferred. Note the center tap on the secondary of this transformer. No center tap is required on previous circuits so some readers might miss this essential connection!

The rectifier is a bridge stack that will pass at least 2 amperes without harm. *Note*: 3-amperes is a standard size, as is 1.5 amperes. The latter, however, will require de-rating of the overall output current rating of the power supply.

This power supply uses the transformer center tap as the zero potential reference for the power supply. Therefore, it is connected to the common line. The transformer is used such that one-half of the secondary supplies the positive side of the supply and the other half supplies the negative half. The rectifier is used in a manner that can be described as a pair of interconnected half-wave bridge rectifiers.

The filter capacitor is selected according to the 2000 μF/

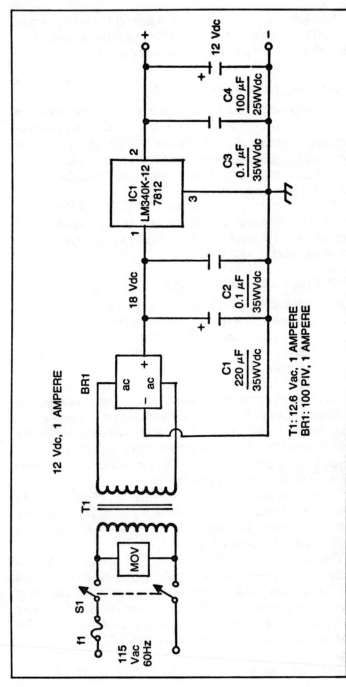

Fig. B-2. A 12-volt, 1-ampere power supply.

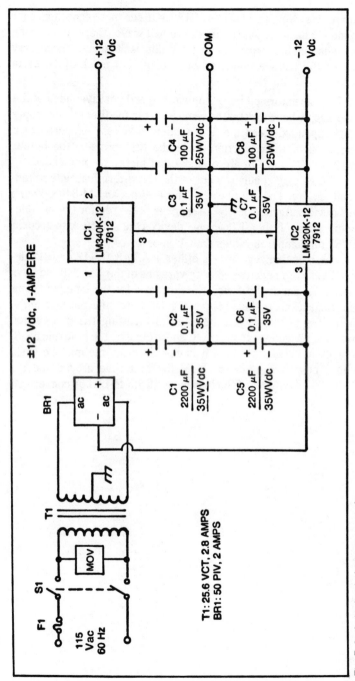

Fig. B-3. A dual-polarity power supply.

417

ampere rule. Note that purists will want to see more capacitance for C1 and C5 because the rectifiers are half wave (thereby requiring somewhat higher capacitance). But this isn't strictly necessary because the voltage regulator will smooth out much of the extra ripple.

The capacitors on both the negative and positive sides of the power supply have exactly the same use as in the first power supply in this appendix. Read the first section (5-volts, 1-ampere) for a description of their function. Like the first project, the output capacitors (C4 and C8) are optional, but highly recommended.

The voltage regulators are of the three-terminal, integrated-circuit variety. The positive regulator is either an LM-340K-12 or a 7812 device. Again, the "K" package is preferred for the full one-ampere rating. If the "T" package devices are selected, then count on 750 milliamperes rather than 1 ampere.

The negative regulator is either an LM-320K-12 or 7912 device. These are negative polarity versions of the LM-340 and 78xx devices, respectively. *Note*: The terminals for the input and common connections to the negative regulator are different than in the positive regulator. If you are used to inputting the unregulated power supply voltage to pin no. 1, and grounding the case (pin no. 3) on positive regulators, then it is easy to make the mistake when using a negative regulator. I did the first time around and the LM-320-12 went *pooofff* and became a silicon-to-carbon converter!

Index

Index

421

OTHER POPULAR TAB BOOKS OF INTEREST

TAB   TAB BOOKS Inc.

Blue Ridge Summit, Pa. 17214

Send for FREE TAB Catalog describing over 750 current titles in print.